EATEN BY A
GIANT
CLAM

EATEN BY A
GIANT
CLAM

Great adventures in natural science

JOSEPH CUMMINS

PIER **9**

CONTENTS

INTRODUCTION

At one time, right up until the mid-twentieth century, being eaten by a giant clam seemed one of the most exotic fates that could befall any adventurer in the natural world. It meant, in a sense, that you had arrived, had earned your stripes as a true explorer of the wild— posthumously, of course. People simply accepted the fact that being devoured by a giant mollusc was a possibility.

A 1924 issue of the famed magazine *Popular Mechanics* displayed a murky picture of one evil-looking enormous bivalve and wrote: 'Divers who often step into the open lips of such monsters are frequently held with such force that they cannot release themselves and drown. The shells ... serve as gigantic traps.' During World War II, the official *US Navy Diving Manual* gave detailed advice to its frogmen on how to cut a giant clam's abductor muscles (used to open and close the shell) if trapped. 'The Attack of the Giant Clam' was once a staple of cinematic art. Early movies featured underwater scenes of killer bivalves, not just catching and holding humans, usually pearl divers, but swallowing them whole. The 1948 movie *16 Fathoms Deep* had a scene where a diver was caught by a giant clam and drowned—off the coast of the southeastern United States, no less (the giant clam is native to the warm seas of the Indo-Pacific region, not the Atlantic).

But then, alas, debunking in the form of cold, hard scientific fact set in. 'Although there are no accurate reports of people being caught by giant clams, their reputation is extremely bad. Popular legends describe giant clams trapping men with an incredibly powerful vice-like grip and eventually drowning them', wrote Ben Cropp in his 1986 book *Dangerous Australians: The Complete Guide to Australia's Deadliest Creatures*, published by no less an authority than Murdoch Books. 'However, as the giant clam is both conspicuous and slow to close its valves, the danger is minimal.'

The giant clam—a truly fascinating creature that can grow up to 1.2 metres (4 feet) across, weigh over 200 kilograms (440 pounds) and enjoy an average life span of a hundred years or more—is, in reality, nature's couch potato, in adulthood being entirely sessile, meaning that it simply cannot move. To be caught between the lips of a giant clam one would have to float down, stick one's leg inside and then settle down to read a few chapters of *War and Peace* before one would feel even the first tender nibble.

There are, therefore, no stories of people being eaten by giant clams in *Eaten by a Giant Clam: Great adventures in natural science*, for the very simple reason that it is unlikely that this ever occurred. But the title is apt because it shows what these naturalists risked—because they lived in a world where such a thing was thought entirely possible. And even stranger fates might await them in the wild. Most of the stories in the book (which spans 450 years) come from that golden age of natural adventuring, the nineteenth century. Walking into the wild, protected by little more than the power of their dreams and aspirations, adventuring naturalists punched pythons in the mouth, attacked bats with violins, crawled into tiger's dens and fought off attacks by spiders the size of birds. This is not to mention the problems they encountered with that far more dangerous species, humankind. In these pages, you will find a British scientist grappling with a cannibal in the wilds of northern Canada, a Scottish flower hunter escaping vicious attackers in Tibet, a daring American palaeontologist shooting it out with bandits in Outer Mongolia, and a German prince watching a horrifying slaughter of Indians in America's Far West.

This is a book not about scientific theory, but about science in practice. A great deal of what we know about the natural world is because of the daring of the men and women we meet inside these pages. The pioneering Dutch naturalist-painter Maria Sybilla Merian travelled to Surinam in the late seventeenth century to capture the brilliance of metamorphosis on her canvases. Along with James Cook, Joseph Banks made brilliant discoveries in the South Pacific and Australia that changed the European conception of the world. Flower and plant hunters such as David Douglas, George Forrest and Joseph Hooker travelled the globe, making discoveries that flower today in our gardens and parks. The Amazonian explorer Henry Walter Bates—who had a very close encounter with some curl-crested toucans—brought back thousands of new species from the jungles of South America, showing in practice how Darwin's theory of evolution worked.

The naturalists in this book were, I suspect, committed to the adrenaline rush many people now get, in an age of dwindling natural frontiers, from adventure sports. They took chances. They went back into the wild over and over again, despite repeated brushes with death. None of them was eaten by a giant clam, but if becoming a meal for a monstrous mollusc were possible, they might have arranged it. For science's sake, of course.

But secretly for the thrill of it.

THE JOHN TRADESCANTS
'what is rare in land, in sea, in air'

A love of nature often runs through families—one thinks of William Bartram and his father (see page 66)—but one of the prime father and son teams in all natural history is that of John Tradescant, *père et fils*, who are so coupled that they are often spoken of in the plural: the John Tradescants. Spanning the period from the Elizabethan era to the turbulent time of King Charles I, the two men were gardeners to royalty, introducing numerous new plant species to the British Isles. They travelled far into the Old and New Worlds (John the Elder to Russia, the Near East and North Africa; John the Younger to the burgeoning new colony of Virginia) and created a collection of curiosities—known as the 'Ark'—that formed the basis for England's first museum.

The two, in fact, are so linked that they are buried beneath a single tombstone whose epitaph celebrates them as men who 'Liv'd till they had travelled Orb and Nature through' and discovered 'what is rare in land, in sea, in air'.

'Wheare the Rarest thing wear'

John Tradescant the Elder was born, probably around 1570, to a yeoman named Thomas Tradescant in the Suffolk coastal village of Corton. Not a great deal is known about either his family or his early personal life, or how he learned his trade as a gardener. However, he must have been quite accomplished, for by 1610 he had a powerful patron: Robert Cecil, the first earl of Salisbury, Secretary of State to Queen Elizabeth I and her successor after 1603, King James I, but also England's top spymaster. The brilliant Cecil, a tiny man with a pronounced hump, owned Hatfield House in Hatfield, Hertfordshire, which included two parks, a moat, a vineyard, a bowling alley and numerous fish ponds, and also boasted 17 hectares (42 acres) of gardens that were, in the main, laid out by John Tradescant.

It was because of Cecil that we have our first clue to John Tradescant's presence in the world, since in order to acquire plants and shrubs from the Low Countries for Hatfield's gardens, Tradescant initiated a series of letters to an English diplomat named William Trumbull, who helped grease the wheels for him in Brussels. Tradescant's spelling, even by the standards of a freewheeling era, is something to behold:

> *Since I laste was with yr Worshipe [Trumbull], it hathe plessed my Lord tresorur [Lord Cecil] to give me eterteynment and he Spake to me to know wheare the Rarest thing wear then I tould him of Brussell then my Lord said Sirra Remember me and at miccalmas yw shall goe over now ...*

Aside from the fact that this indicates that Tradescant's education was in gardening, not the English language, the letter shows that he had a good deal of knowledge about where unusual botanical prizes could be found. He sent Trumbull a detailed list of grapevines to acquire. Hatfield House's kitchen gardener, Tradescant wrote, wanted 'the blewe muskadell the Russet grape a greatest quantity of those but of aull other sorts', as well as flowers of every stripe—paradise lilies, irises, star-of-Bethlehems.

In late 1611, Cecil sent Tradescant in person to the Low Countries and France in order to purchase plants. Tradescant bought a staggering amount in what is now the Netherlands—800 tulip trees, 200 lime trees (which would eventually line the great drive leading up to Hatfield House), thirty-two cherry trees and two mulberry trees. Back at home early in 1612, Tradescant would have been busy at work setting up the gardens and spending time with his wife, Elizabeth, of whom not much is known, and his four-year-old son, John. Such family time was probably scarce, however, since he was also helping design the gardens at Cecil's London home, Salisbury House.

'A viag of Ambasad'

Robert Cecil died of illness in May of 1612 but John Tradescant stayed on two more years, working for Cecil's son, William, before moving on to design gardens for Edward, Lord Wotton, at St Augustine's Abbey in Canterbury, a post nowhere near as distinguished as that of gardener at Hatfield House but one filled with interesting challenges nonetheless. Wotton gave Tradescant a free hand with his gardens, and Tradescant responded by ambitiously filling them with unusual and rare plants—a male mandrake, a new kind of pomegranate and an assortment of wild garlic plants. While Tradescant did not go to the new colony of Virginia himself (as his son eventually would) he was friendly with Captain Samuel Argall, a sea captain employed by the Virginia Company, who would go on to become deputy governor of the colony (and be the one who held the famous Pocahontas for ransom). Tradescant invested twenty-five pounds—half a year's salary for him, so quite a large sum—in the Virginia Company and may have reaped the benefits of the tobacco crops beginning to be exported to England. He definitely received plants for Wotton's garden, including one, the Virginia stinkwort, which has become known as Tradescant's stinkwort, or *Tradescantia virginiana*.

But Tradescant's biggest adventure for Lord Wotton was a trip to Russia, which he began in June 1618. At the time, Russia was engaged in a war with Poland, and Czar Mikhail Fyodorovich, the first of the Romanovs, wanted to strike an alliance with England—in particular, he desired a large loan in order to buy guns and ammunition. The English finally proposed to lend him sixty thousand pounds in return for his letting English merchants cross Russian territory to Persia without hindrance.

King James I chose Sir Dudley Digges to sail to Archangel, taking with him twenty thousand pounds to seal the deal. There were seven ships in his fleet, two of them carrying Russian envoys back to their country. Digges took forty-one men along with him and one of them was John Tradescant. This was a fact unknown to history until two centuries later, when an anonymous twelve-page manuscript in Oxford's Bodleian Library—entitled 'A viag of Ambasad [A Voyage of Embassy] undertaken by the Right honnorabl Sr Dudley Diggs In the year 1618'—was identified by scholars as having been written by John Tradescant. As Tradescant scholar Jennifer Potter says: 'It is the only sustained piece of Tradescant writing known to have survived.'

Tradescant probably was invited along on the voyage through the intervention of Lord Wotton, although this is not certain. He did not accompany the mission in any official capacity, although he was certainly more than a tourist. Most likely, he was there to seek out new botanical wonders and to make a report back to Wotton as to the state of trade with Russia. In any event, Digges's little fleet set sail from Gravesend on 3 June and travelled up the North Sea coast of England, facing terrible weather: 'All our landmen fell sick, and my Lord himselfe for 4 daies very sick.'

As the voyage went on, the wind moderated and Tradescant noted whales cavorting near the ships and numerous sea birds. To commemorate the favourite Russian holy day of St Peter, Tradescant observed that Digges sent to the Russians a present consisting of 'on quarter of mutton, half a littill porker, and 3 live pullet, ther Lent being but then ended. Also ... my Lord sent hime two small salmons and 9 gallons of Carnary Sack'. Tradescant then sniffed disapprovingly—hinting at some enmity with the Russians: 'The curtiseys hathe pased a yet witheout requittall.'

'Single rosses wondros sweet'

It took Tradescant and the fleet forty-three days to make it to Archangel, travelling along the coast of bleak northern regions such as Lapland, about which Tradescant wrote:

> ... *if I mought have the wholl kingdom to be bound to live ther, I had rather be a porter in London, for the snow is never of the ground wholly, but liethe in great packes conttinnewally. Ther is no shadowe for the sun shinethe ther continnewally when it is no foggs whiche most tims it is.*

Arriving in Archangel on 16 July, Digges, with Tradescant alongside him, entertained Russian dignitaries in the cabin of his flagship. He also entertained some curious nomads from northern Siberia. These were, Tradescant wrote:

> ... *a misserable people of small grouth. In my judgment is that people whom the fixtion is fayned of that should have no heads, for they have short necks and commonly wear ther clothes over head and shoulders. They use boues and arrowes. The men and women be hardit knowne on from the other because they all wear clothes like mene and be all clad in skins of beasts packed very curousile together, stokins and all. They kill most of the Lothi deer that the hids be brought. They be extreme beggars not to be denied.*

Tradescant had a better time of things when 'on of the Emperors boats' (apparently, a boat belonging to the emissaries of the czar) took him on a tour of some of the islands off the Russian coast 'to see what things growe upon them'. He observed 'single rosses wondros sweet withe many other things which I meane to bringe withe me'.

By 22 September, Tradescant had returned to England, thanking God that not a single person had been lost during the journey. It was not, however, much of a diplomatic success. While Tradescant lingered along the coast, Digges had set out for Moscow but, apparently feeling nervous about travelling through the impending Russian winter with so much money, abruptly turned back to Archangel, leaving the money behind with other emissaries. By the time these made it to Moscow, Czar Mikhail suspiciously refused to accept the loan, not understanding why Digges had turned back. He finally did take the twenty thousand pounds in March 1619, although by this time Russia and Poland were at peace. It appears that the czar never repaid the money, nor did he extend Persian trading rights to English merchants.

The journey was successful for John Tradescant, however, broadening his horizons and making him even more employable.

John Tradescant at war

As if the trip to Archangel put tame collecting visits to the Low Countries to shame, John Tradescant continued to lead an ever more adventurous life. King James I hated the Barbary pirates preying on English shipping in the Mediterranean and when, in 1620, he sent a war fleet to Algiers to destroy the pirates' base and free English captives, Tradescant, at the behest once again of Lord Wotton, joined the fleet. The fleet, under an inept admiral, was notably unsuccessful in its task, but Tradescant, as he did everywhere he went, managed to scour the countryside for treasures, bringing back a wild pomegranate tree whose flowers, according to a contemporary, were 'farre more beautiful than those of the tame or manured sort'. Tradescant also brought back to England Spanish onions (the fleet had stopped in Spain en route) and numerous wild Mediterranean flowers.

Tradescant had been chief gardener to Lord Wotton in Canterbury for some time but a change in his employment now occurred, perhaps occasioned by the fact that Wotton seems to have fallen out of royal favour (he would die in 1626). When next Tradescant appears in English records, he has landed on his feet in a very big way. He is now chief gardener to George Villiers, the first duke of Buckingham and Lord Admiral of England. Villiers—tall, handsome, polished—had been born the son of a minor gentleman, but his good looks and sophistication brought him, in his early twenties, to the eye of King James, who did not care much for women. The two men became very close and Villiers may have been one of the king's 'favourites'—those with whom the relationship turned sexual. Villiers referred to the king as 'Dear Dad', while James called his young friend 'my sweet child and wife'.

Whatever the exact nature of their bond, Villiers rose quickly in the country. Within ten years of meeting James in 1614, he had become a duke, Lord Admiral and the second most powerful person in England. He wielded his power ruthlessly, a tendency for which he was much hated, but if an ambitious gardener could choose any man besides the king to work for, it would be Buckingham, with his vast estates. This made Tradescant the most prominent gardener in England, although it also exposed him to some unusual risks, as when in 1627, aged fifty-seven, he was forced to accompany the duke on a bloody expedition to support Huguenots besieged by French Catholics at La Rochelle. Out of seven thousand men who set off, only three thousand returned. The following year Buckingham was assassinated by a disgruntled officer who had been wounded on the expedition.

The cabinet of curiosities

Once again, John Tradescant landed on his feet, and in 1630 he was working directly for Charles I, who had become king of England when James I died in 1625. And now Tradescant was also known for something more than gardening. During all his travels, he had been collecting—or having collected for him—numerous objects that were generally known as 'curiosities'. He wrote that he had acquired, or wanted to acquire, such objects as the head of a sea cow (a manatee), the skeleton of a river horse (a hippopotamus) and, as he wrote to a friend, 'the Greatest sorts of Shellfishes Shelles of Great flying fishes & Sucking fishes with what els strang'.

And it was not just animal remains Tradescant wanted—he would gather pieces of the true Cross, flutes and weapons from the coast of West Africa, a hand that had supposedly belonged to a mermaid, and much more. He said that he was collecting all this for the duke of Buckingham, but one senses his absolute fascination with the material himself. What he was doing was not unprecedented—so-called cabinets of curiosities, containing all the marvels of the world, were favourites with gentlemen collectors in England and all over Europe.

But, when he got back from fighting in France, Tradescant decided to do something very different—to put all his collected rarities in one place and allow the public to see them. This place would be in South Lambeth, London. When Tradescant moved there in the late 1620s with his family, it was still quite rural, giving Tradescant and his son John—who was now helping with his father's business—the chance to plant a huge botanical garden. By 1632 the garden contained nineteen plants that Tradescant had introduced to England from North America, the Mediterranean and Western Europe. These included the Canada goldenrod, a primrose-yellow sunflower, a kin to the black-eyed Susan, several rockroses and, unfortunately, poison ivy (one of Tradescant's rare miscues).

Among the common people of London, however, Tradescant and his son became much more famous for their cabinet of curiosities, which most people called 'Tradescant's Ark', because it contained all varieties of things. The building that contained the Ark was reached by walking through an arch made of whalebones into a courtyard, and from there to the upper floor of an old house. There everything lay in a jumble, crowded together, and visitors needed the help of

Tradescant himself—a bearded, dignified figure—to tell them just what each item represented. Soon, in order to defray expenses, Tradescant charged a small amount for people to view his curiosities, and by his death in 1638 Tradescant's Ark was on its way to becoming the world's first public museum.

Chief Powhatan's Mantle

John Tradescant the Elder was a tough act to follow. At the time of his death, in April 1638, he was running both his botanical garden and the Ark, was chief gardener to Queen Henrietta Maria, wife of Charles I, and had begun working with the University of Oxford on a 'physick' garden—a garden growing plants to heal the sick. His son, John Tradescant the Younger, shared his love of gardening but was a very different person.

Tradescant the Younger, born in 1608, went to the King's School in Canterbury until he was fourteen, at which time, like most young men, he was apprenticed to a profession—although in this case it was to his father, whose gardening he had almost certainly been helping for years anyway. At the age of twenty, in 1628, he married a woman named Jane Hurte, by whom he had two children, a boy and a girl. In 1634, he joined a guild, the Company of Gardeners—his father did not belong to it, but most gardeners in London did—and the following year, after his wife had died, he moved into his father's house with his two children. (The elder Tradescant was by this time a widower as well.)

When the elder Tradescant died in 1638, his son took over his duties as chief gardener to Queen Henrietta Maria and performed admirably, but a contemporary, a gardener named Walter Stonehouse who knew both father and son, wrote that the young Tradescant was a quite modest man, one who felt that he was labouring in the prodigious shadow cast by his father. In order perhaps to escape this shadow, he did something his father had never done—journeyed to the colony of Virginia in the New World, at least once and possibly three times between 1637 and 1662. (He had just returned from a trip to the New World when his father died.) Tradescant's journeys were quite literally fruitful. He introduced to English gardens great American trees such as magnolias, the bald cypress and the tulip tree, and brought back such plants as the Virginia creeper bush, the runner plant and the scarlet bean.

He did not forget the Ark, which continued to thrive and grow after the death of his father. He returned from the New World with such prizes as 'Indian Crownes made up divers sorts of feathers', 'Virginian purses imbroidered with Roanoake', six different kinds of 'Tamahacks' (tomahawks) and a few stuffed rattlesnakes, to boot. But the greatest stroke of fortune for the Ark was the cloak that was called Chief Powhatan's Mantle. Chief Powhatan was, when the English arrived in Virginia, the most powerful chieftain in the area, and the father of Pocahontas. The cloak was made from four carefully prepared pieces of tanned buckskin and decorated with seventeen thousand marine shell beads, which were arranged to show a human figure flanked by two deer or bears. It is still in the Ashmolean Museum at Oxford, and although it is not universally recognised that it belonged to Powhatan, it was certainly the property of a very high-ranking Native American.

A rare ark indeed

During the course of their travels, both Tradescants, but especially John the Elder, brought home any number of rarities to their famous 'Ark' in South Lambeth, London, a private collection that soon grew into a museum open to the public. In 1638, the year of John Tradescant the Elder's death, a German traveller named Georg Cristoph Stirn gave this breathless record—in a single, very long sentence—of the rarities the Ark contained:

In the museum of Mr. John Tradescant are the following things: first in the courtyard there lie two ribs of a whale, also a very ingenious little boat of bark; then in the garden all kinds of foreign plants, which are to be found in a special little book which Mr. Tradescant has had printed about them. In the museum itself we saw a salamander, a chameleon, a pelican, a remora, a lanhado from Africa, a white partridge, a goose which has grown in Scotland on a tree, a flying squirrel, another squirrel like a fish, all kinds of bright coloured birds from India, a number of things changed into stone, amongst others a piece of human flesh on a bone, gourds, olives, a piece of wood, an ape's head, a cheese, etc; all kinds of shells, the hand of a mermaid, the hand of a mummy, a very natural wax hand under glass, all kinds of precious stones, coins, a picture wrought in feathers ... two cups of rinocerode, a cup of an E. Indian alcedo which is a kind of unicorn, many Turkish and other foreign shoes and boots, a sea parrot, a toad-fish, an elk's hoof with three claws, a bat as large as a pigeon, a human bone weighing 42 lbs., Indian arrows such as are used by the executioners in the West Indies—when a man is condemned to death, they lay open his back with them and he dies of it, an instrument used by the Jews in circumcision, some very light wood from Africa, the robe of the King of Virginia, a few goblets of agate, a girdle such as the Turks wear in Jerusalem, the passion of Christ carved very daintily on a plumstone, a large magnet stone, a S. Francis in wax under glass, as also a S. Jerome, the Pater Noster of Pope Gregory XV, pipes from the East and West Indies, a stone found in the West Indies in the water, whereon are graven Jesus, Mary and Joseph, a beautiful present from the Duke of Buckingham, which was of gold and diamonds affixed to a feather by which the four elements were signified, Isidor's MS of de natura hominis, a scourge with which Charles V is said to have scourged himself, a hat band of snake bones.

'The Musaeum Tradescantianum'

It will never be known whether Tradescant the Younger could have equalled his father's great career, for the outbreak of the English Civil War, which pitted Charles I against Parliament and the Puritans of Oliver Cromwell, definitely curtailed his work. Queen Henrietta Maria was forced to flee for her life in 1642, and it is doubtful whether Tradescant would have continued to be paid as her gardener. By this time he was married again, to one Hester Pooks, and the lack of income from the royal family made things quite difficult.

Even more difficult was the fact that, like his father, Tradescant was a royalist, and London, where he lived, was a city that supported Parliament. When Charles I was executed in 1649, Tradescant lost any hope of working for the royal family, but he apparently continued working in his botanical gardens, where he did well with his orchids and grew for sale apple, pear, cherry and plum trees—more useful plants for an age that was beginning to eschew the exotic 'curiosities' so beloved of Tradescant the Elder and his contemporaries.

Tradescant did not neglect his father's museum of oddities either, expanding and, by 1652, cataloguing the jumbled collection, at the suggestion and with the help of a lawyer, Elias Ashmole, who had befriended Tradescant and his wife. Tradescant renamed the collection 'the Musaeum Tradescantianum' and it continued to draw crowds paying a small price to gape at the wonders of the world and nature.

John Tradescant the Younger died on 22 April 1662, at the age of fifty-three. Although by this time Charles II had gained the throne of England, Tradescant had not gone back to work for English royalty but seemed content to work in his orchards and develop the museum. Unfortunately, a vicious dispute with Elias Ashmole over ownership of the Tradescant collection would forever remove it from the Tradescant family, but Tradescant's legacy literally grows on in England, as does his father's, in all the many species of plants and trees that the two men introduced to their country—'curiosities' that have now become quite familiar.

The death of Hester Tradescant

Elias Ashmole, born in 1617, was a lawyer and an antiquarian expert, who met John Tradescant the Younger around 1650. Ashmole was fascinated by the Tradescants' collection of rarities and convinced Tradescant that he should catalogue it—not only would Ashmole help him, but Ashmole would also pay for the publication of the catalogue listing the contents of the Musaeum Tradescantianum.

All of this was an offer that perhaps Tradescant the Younger should have examined more closely, but at the time he felt he needed all the help he could get, and so he let Ashmole into his life. In 1659 Ashmole convinced Tradescant to deed him ownership of the museum after his death—Hester Pooks Tradescant, Tradescant's wife, was certain that Tradescant had been drunk when he did this. When Tradescant died in 1662, he left all his property to Hester, with the exception of the collection of rarities. Hester, who had not known that John had turned over the collection, was outraged and fought Ashmole in court. There ensued a vicious two-year-long legal battle, which ended with Hester losing the collection—although the courts allowed her to keep it in her house until her death—and being forced to apologise humiliatingly for attacks on Ashmole. To make matters worse, Ashmole moved into the house next door to the Tradescants' in order to keep an eye on his collection.

After some years of this humiliating existence, Hester drowned in the pond on the Tradescant property. Ashmole writes laconically in his diary entry of 4 April 1678: '11:30 AM. My wife told me that Mrs. Tradescant was found drowned in her Pond. She was drowned the day before about noone as appeared by some Circumstances.'

No one will ever know, but it appears that Hester committed suicide, although her death was officially ruled an accident, which allowed her to be buried in sanctified ground. Ashmole moved immediately to acquire the collection of rarities, placing it in his own house and closing it to the public, allowing only dignitaries to view it and calling it 'the Ashmolean Collection'. Interestingly, Ashmole's perfidy actually preserved the collection for posterity, for he eventually made a gift of it to the University of Oxford, where the Ashmolean Museum is one of the finest in the world today.

MARIA SIBYLLA MERIAN
The wonderful
transformation

Metamorphosis—that process of transformation by which a living being changes in form and nature—is at the centre of the symbolic thinking that has moved artists, writers, great religious and political thinkers, and even nation-states. To transform completely, to shed one's old self and emerge anew, is such a resonant and magical concept that many of us try all our lives to attain some semblance of it, although it is death that is often portrayed as the most powerful of metamorphoses.

Of course, metamorphosis begins in the natural world, notably when a caterpillar transforms in stages into a moth or butterfly. In the late seventeenth and early eighteenth centuries, naturalists were fascinated by this process, but no naturalist of the period was as obsessed by it as Maria Sibylla Merian. A brilliant artist whose paintings of insects and plants are luminous even hundreds of years later, an equally brilliant naturalist, and a woman in a man's world, who shed a husband and travelled thousands of kilometres at a time when most women her age were dandling grandchildren upon their knees, Merian's very life is an extraordinary example of human metamorphosis at its finest.

A childhood in art

Maria Sibylla Merian was born in Frankfurt am Main on 2 April 1647 into a world transformed by the violence and chaos of the Thirty Years' War, which would come grinding to a halt in 1648, having decimated most of Europe. The Thirty Years' War was a religious conflict unprecedented in its scale and brutality. Germany, where Merian was born, lost between a third and a half of its population. Yet she was fortunate to be born in Frankfurt. It was a city largely protected from the worst of the war's excesses (although fully a third of the city's twenty-one thousand inhabitants had died in a bubonic plague outbreak in 1636) because it was a powerful commercial centre, situated as it was on the Main River, a major trade route feeding into the Rhine. It was also a city that welcomed and sheltered people of different religious beliefs, so that Protestants, Catholics and, to some extent, Jews lived together in relative peace and quiet.

Merian was the daughter of artist and publisher Matthäus Merian the Elder and his second wife, Johanna Sibylla Heim. Matthäus ran a lucrative publishing house known for its large-format picture books and maps. He illustrated many of these books, as did the older children from his first marriage, Matthäus the Younger and Caspar. Matthäus the Elder died when Maria was only three years old, and within a year her mother remarried, to a still-life painter named Jacob Marrel.

All of this is to say that, given her father, half-brothers and stepfather, Merian was exposed to art at a very young age. As the Thirty Years' War receded into the past, Frankfurt came to the fore as a cultural centre, even then holding yearly book fairs and hosting numerous rich artistic communities. While it may not be true, as possibly apocryphal tales have it, that Merian was fascinated at an early age by the caterpillars and other crawling creatures in the family garden, it is certainly true that she had a distinct advantage when it came to learning art. Her stepfather, Jacob Marrel, took on numerous apprentices, and she was able to sit in on classes he gave in watercolours, engraving and still-life painting (although the painter's guild would not allow women to paint in oils, which were exclusively, they felt, a male province). Maria would not have been allowed to attend the life-study classes (with their nude male models), nor would she have been able to undertake a seasoning trip abroad, as her half-brothers did. But she was trained, and because of her family background she received a far more intensive training in the craft of making art than most other girls of her time.

'They make one alive through the other'

Jacob Marrel specialised in painting flowers, so it is only natural that Maria would also gravitate in that direction. Insects were often used as motifs in these paintings (sometimes to symbolise virtue, sometimes evil) and Marrel would send Maria out into the garden to find insects to pin to boards and use as models. Maria left few writings behind that did not directly relate to her art or passion for nature, but she did relate in one of her first books that she was encouraged to do her plant drawings 'decorated with caterpillars and summer birds and such little animals, like the landscape painters do. They make one alive through the other.'

Unfortunately, the family began to suffer upheavals. The two children that Johanna Heim had with Jacob Marrel, a boy and a girl, both died very young and Marrel finally decided to leave Johanna and Maria and return to his native Netherlands in 1659, when Maria was twelve. Essentially, it was the end of the marriage, although he did provide financial support and occasionally visited. With her half-brothers gone and her mother no doubt in a state of shock, Maria turned for solace to studying insects, including caterpillars and silkworms: 'I realized that caterpillars produced beautiful butterflies or moths, and that silk worms did the same. This led me to collect as many caterpillars as I could find in order to see how they changed.'

Already, Maria Merian was ahead of many of her time. It was thought that insects arose from the earth or dung or garbage due to abiogenesis (spontaneous generation) or were somehow born in drops of dew. But with the invention of the microscope, which allowed for close observation of tiny insects, and with a new breed of natural scientists who did not allow themselves to be swayed by tradition or religion, these beliefs began slowly to change. In 1662 a Dutch painter and natural scientist named Johannes Goedaert began publishing a series of three volumes he entitled *Metamorphosis Naturalis*, in which he closely observed and sketched caterpillar transformations.

Merian found a copy of *Metamorphosis Naturalis* and read it avidly. Goedaert was nowhere near the painter she was (and he hedged his bets by avowing a belief in spontaneous genration in some cases) but, im Todd, one of Merian's recent biographers, has written, the most important thing she learned from the Dutch artist was his methods. He did not kill the caterpillars, but 'kept them in jars, fed them leaves, and watched their changes through a microscope ... noting details of how they transformed and then painting what he saw'.

The Wonderful Transformation

In 1665, at the age of eighteen, Maria married a former apprentice of Jacob Marrel, a twenty-eight-year-old painter named Johann Andreas Graff. In 1668 they had a daughter, Johanna, and in 1670 they moved to the city of Nuremberg, where Graff had been born. He busied himself with detailed but mainly pedestrian cityscapes, and Maria—while taking care of the household and raising her daughter—continued to paint flowers and insects, especially silkworms, and took in students to help defray the family expenses. By 1675 she published her first book, *Blumenbuch*, literally, a 'book of flowers', which depicted colourful blossoms in wreaths and bouquets and which she sold in loose bunches of twelve pages. There was no prose commentary beyond the names of the flowers, but there was the occasional insect—a spider or caterpillar or butterfly—on the blossoms themselves. People commented on the verisimilitude of the plants and insects and the book was a hit in Nuremberg. Maria (who had published *Blumenbuch* under the name 'Maria Sibylla Graffin, daughter of Matthäus Merian the Elder') began to receive public praise for her work, something that may have been difficult for her husband's ego. Another former apprentice of her father, Joachim von Sandart, in writing of talented German artists of the year the *Blumenbuch* appeared, listed Maria enthusiastically as being known for:

> ... *all kinds of decorations composed of flowers, fruit and birds, and in particular also the excrement of worms, flies, gnats, spiders and all such kind of creatures with their possible permutations; she showed how each species is conceived and subsequently matures into a living creature.*

Although this estimation is flattering, it still focuses on Maria as a woman (Sandart makes sure to note that she was a wonderful housekeeper!). But it is obvious, as Sandart knew, that she had more in mind than simply painting flowers. In 1679 she published her first great work, entitled *The Wonderful Transformation and Singular Flower-Food of Caterpillars ... Painted from Life and Engraved in Copper.* A second volume followed in 1683, and one more would be published posthumously in 1717. Each book had fifty plates and came in two versions, one hand-coloured for the buyer who could afford it, and the other in black and white.

The Wonderful Transformation is a book of art, but it is also a book of natural science. Each of the plates showed a plant that is food for caterpillars, usually in its flowering stage, and then what Kim Todd calls 'a drama of miniature proportions'. You see the caterpillar, its pupa, or cocoon, the butterfly, the butterfly laying eggs and sometimes the caterpillar's predator (usually a fly). An entire cycle of life and death is represented—a true saga of transformation.

'He will indeed need good advice'
At about the time the second volume of *The Wonderful Transformation* appeared, Maria Sibylla Merian was undergoing a metamorphosis herself. In 1681 her stepfather, Jacob Marrel, died, and Maria, her

two daughters (a second daughter, Dorothea, had been born to her in 1678) and her husband journeyed back to Frankfurt to be with her mother, to whom Maria was extremely close. There was family squabbling involving Marrel's will and estate, which was not settled until the summer of 1685, but, when it was, Johann Graff returned to Nuremberg. Maria and her two daughters were not with him and did not intend to join him. There had been no hint of discord in the marriage in the letters Maria wrote (those that are extant, anyway), but rumours of Johann's infidelity have come down over the centuries. However, what was troubling the marriage is not known with any certainty. That summer of 1685 Maria wrote to her friend and former student Clara, back in Nuremberg: 'I have no news, aside from the fact that my husband wants to journey to Nuremberg ... I ask that if he should need counsel you should be recommended to his humble person, for he will indeed need good advice.'

Possibly Maria's burgeoning fame disturbed Johann, or possibly he was unfaithful. But a third possibility is that Maria rejected her husband (or he turned his back on her) because of her growing attraction to a religious sect known as the Labadists, founded by a former French Catholic, Jean de Labadie, some twenty years before. Maria had been raised a Lutheran, but her study of nature had deepened her faith and made her unhappy with conventional religion. Metamorphosis, she felt, happened through the grace of God. In the preface to the first volume of *The Wonderful Transformation*, Maria wrote:

> *These wondrous transformations have happened so many times that one is full of praise for God's mysterious power and his wonderful attention to such insignificant little creatures and unworthy flying things ... Thus I am moved to present God's miracles such as these to the world in a little book. But do not praise and honor me for it; praise God alone, glorifying Him ...*

With these intense religious feelings, it was probably only a matter of time before Maria needed to find an outlet in a faith more ecstatic than conventional Lutheranism. The Labadists—whom she very likely heard about through her half-brother, Caspar, who was a convert— fit the bill nicely. Although Jean de Labadie had died in 1674, he had preached a faith stripped down to its bare essentials, without the hypocrisy and corruption of organised religion and, indeed, society at large. In order to have 'true Christian friendship' Labadists retired from society into communities that promised what Labadie had called a 'Paradise on earth', stripped of social ranks, where praying and contributing to the communal good were the highest callings. One of the chief Labadist communities—where, in fact, Caspar was living in 1685—was in Wiewert, in Friesland, north of Amsterdam. There some 350 Labadists lived in and around a large manor house known as Waltha Castle, formerly owned by an aristocratic family who had given up everything years before for Jean de Labadie (the sixty-one- year-old guru had repaid them by marrying their twenty-two-year- old daughter, nearly concurrently with impregnating several young women who came to Waltha, one reason why a celibacy requirement for residents there was soon dropped).

Shortly after her husband departed for Nuremberg, Maria Sibylla Merian, her two daughters and her mother departed for Waltha Castle to join the sect.

The Star of Utrecht

Probably the most famous member of the community at Waltha Castle was Anna Maria van Schurman, a writer born in Cologne in 1609. Like Maria Sibylla Merian, she was an amazingly accomplished woman in an era that did not value accomplished women.

Schurman, the daughter of wealthy and educated parents, learned to read when she was only four years old. After her father died in 1613, she moved to Utrecht with her mother and two aunts, becoming the first female student at the university there (although she was forced to sit in a recess, behind a screen, so that her presence would not disturb the male students). Eventually, she learned fourteen languages, including Latin, Greek, Arabic and Ethiopian. In 1639 she published a book called The Learned Maid, or Whether a Maid May Also Be a Scholar, *in which she argued forcefully that women should shun 'narrow limits' and embrace education. Schurman became known as 'the Star of Utrecht' and became an attraction for every scholar or dignitary visiting the town.*

However, in an echo of Maria's experience, Schurman's brother became a disciple of Jean de Labadie, introducing him to Schurman in 1666. Coming under Labadie's sway, she abandoned her life in Utrecht and moved to Amsterdam then Denmark, and finally to Waltha Castle, where she died in 1678, well before Maria's arrival (although Schurman did know Caspar Merian). But Maria undoubtedly felt Schurman's influence, since Schurman was one of the main drawing points of the sect, their star convert, and had written a widely disseminated tract, printed on the Labadist press, in which she renounced her scholarly studies and encouraged other women to do the same.

Behind the high gates

Maria, now thirty-nine years old, was doing an extraordinary thing—leaving her home and husband and striking out on her own, responsible for seven- and seventeen-year-old daughters and an ageing mother. Once this group arrived at Waltha Castle, they shed their comfortable lives. The community was so crowded that each room was crammed full of families. The Merian women would have given up any adornment to wear only the plainest of clothes. Since fuel was an expensive commodity, they were continually cold. People were forced to pool all their money and possessions for the common good, although Maria was able to hang on to her painting tools.

Behind the high gates of Waltha, people gardened and raised livestock, especially sheep. A printing press churned out religious tracts. Children were taught their reading and writing with a focus on Labadist principles. Self-mortification was a large part of these beliefs. Some people practised self-flagellation and children were subject to beatings from the 'aunts' and 'uncles' of the sect. Although the community was supposedly classless, new arrivals were considered 'probationers', who must prove themselves via self-sacrifice and hard work before becoming members of the 'elect', who had greater privileges. Sessions of group criticism were common, with the elect haranguing the probationers.

Maria Sibylla Merian spent six years at Waltha Castle. While many men and women who arrived gave up their callings, Maria continued to work, gathering insects in nearby fields, painting them, dissecting frogs and keeping a record of her close observations in a notebook. The summer after she arrived at Waltha, Johann Graff showed up, wanting his wife and family back. He may have offered to join the Labadists. In any event, he apparently begged her in public to come back, but she refused and their daughters refused to see him. After this he stayed in the town of Wiewert, doing odd jobs, apparently hoping she would change her mind. When she did not, he left and within a few years had filed for divorce.

In 1690 Maria's mother died (her brother Caspar had already passed away). By this time, Maria had apparently grown disillusioned with the Labadist community, if not with its ideals. She was concerned about its practice of banning books inspired by 'the spirit of the world', that is, books of art, poetry, science, the very ideas she had been raised on. While her daughter Dorothea was only twelve, Johanna was now twenty-three and Maria may have had some doubts as to whether she might find a suitable husband at Waltha. In any event, with her mother and brother dead, she decided to break from the community. In the summer of 1691 she left, this time for Amsterdam.

'I lacked the opportunity'

Maria Sibylla Merian's life was already radically different from that of most women of her age. Now forty-five years old, she had had a career as a painter and naturalist, left her husband, entered a religious community and then left the community to strike off on her own in the fourth largest city in Europe. Amsterdam was home to many notable painters and scientists, a huge cultural centre and bustling port, with a population of about two hundred thousand. How were Maria and her daughters to make a living in this teeming city, which must have seemed overwhelming—especially to her daughters—after six years spent in a small commune in the country?

Typically self-possessed, Maria came up with an idea. She had trained both daughters in the art of painting still–lifes, and she decided to have all three of them work in their own studio, producing paintings based on some of the more spectacular images in *The Wonderful Transformation*. (Johanna would go on to become a superlatively talented painter of flowers.) Maria also began to create and sell collections of insects, with each stage in their lives clearly marked. These were quite different from the private collections she had been viewing around Amsterdam, of which she wrote in some exasperation: 'In every collection ... the insects' origin and propagation were absent, that is, how they transformed from caterpillars into pupae and so forth.'

After a few years Maria became quite successful and well known in the scientific and artistic circles of Amsterdam but, with her restless nature, she also became bored. Partly this may have been because of the lack of new and interesting insect specimens for her to find in the large city, for writing of these years some time later she said: 'In Holland more than anywhere else I lacked the opportunity to search specifically for that which is found in the fens and heath.'

She now made an extraordinary decision. One daughter (Johanna) was now married; the other, Dorothea, was twenty-one years old. Her obligations to family were lessening, while her obligation to her art was growing. Even though she was fifty-two years old, Maria decided to leave Holland to go to Surinam, the Dutch colony in South America, in order to find new and exotic breeds of insects to study. She pursued the idea methodically, writing her will, selling off all her stock of paintings to fund the trip, and then, in June of 1699, setting sail with Dorothea for new horizons.

'They are so beautiful'

Surinam was a tiny Dutch colony on the northeastern coast of South America, surrounded by still largely unexplored wilderness. Maria chose to go there in part because it was Dutch, in part because it had a reputation for exotic insects and animals, and in part because the Labadists had formerly had a colony there. She would have been familiar with the area from letters sent back to Waltha Castle and read to the community while she was in residence there.

Maria and Dorothea landed at the small coastal town of Paramaribo in August, found a house and set to work collecting. Maria was thrilled by the wide variety of insects she encountered within a few hundred metres of her house. She plucked poisonous caterpillars from trees, watched as lizards climbed over the walls of her house and carefully captured tarantulas. Of the last, she wrote:

> *They are covered with hair all over and supplied with sharp teeth, with which they give deep and dangerous bites, at the same time injecting a fluid into the wound ... when they fail to find ants they take small birds from their nests and suck all the blood from their bodies. [This last was an observation also made by the nineteenth century Amazonian explorer Henry Walter Bates—see page 213.]*

Gradually, Maria began to move farther afield into the rainforest, whose steamy entanglements made collecting a far more complicated proposition from what it had been in Germany or Holland. She needed the help of local Indians and African slaves to locate and collect plants and insects. Her plan was twofold: to observe and paint insects and animals, and also to collect specimens for rich collectors back in Amsterdam. She netted lantern-flies (Fulgoridae), strange insects with a huge, lizard-like head that she thought (mistakenly) was luminous at night. She collected the Surinam toad, which carries its young in a skin sack on its back. And she continued to paint her staple, caterpillars—green and black ones with yellow faces, which in time became huge ghost moths with wingspans that could reach 30 centimetres (1 foot). She wrote of some of the insects she saw: 'They are so beautiful if one looks at them without the magnifying glass, so strangely ugly, if one regards them with its help.' In some ways, as she got older, Maria wanted to control the experiences she put on the page—she sometimes did not include plants or animals if they were too large to paint readily on a standard sheet of paper.

The ultimate metamorphosis

Sometime in 1701 Maria contracted malaria, which left her weak with fever and chills. Specimen collecting was difficult at the best of times, for people much younger than she was, and even though she recovered from the malaria, the tropical heat began to get to her. Her plan had been to stay in Surinam for five years but, as she wrote: 'The heat in this country is staggering, so that one can do no work at all without great difficulty, and I myself nearly paid for that with my death, which is why I could not stay there longer.'

Carefully packing her paintings and specimens, she headed home in the summer of 1701. If she was unhappy with her forced early departure from the tropics, she did not show it, instead throwing herself into the work that would become *The Insects of Surinam*. Most of the paintings that went into this work existed only as sketches when she left Surinam, and they had to be turned into preliminary watercolours (Dorothea, although she married in December 1701, helped with this) and then into final watercolours. In the meantime, Maria was furiously trying to sell off collections to make money—in 1702 she wrote to a friend in Nuremberg that she had for sale 'animals in liquid', which included one crocodile, two large snakes, eleven iguanas and 'one small turtle'.

Maria Sibylla Merian and Peter the Great

Peter the Great—the modernising, westward-looking czar of Russia—visited Amsterdam in 1717 on a personal tour, along with his personal physician, Robert Areskin, a Scot with an interest in natural history. The czar hired Maria Sibylla Merian's son-in-law, George Gsell (Dorothea's first husband) to act as an art agent, and when Gsell met Areskin he mentioned the number of paintings that existed in the collection of his mother-in-law, who was then very ill. Areskin had heard of Maria and agreed to come to Dorothea and George's home in order to look at the collection with an eye to purchasing it.

Areskin purchased three hundred paintings for the czar at the price of three thousand guilders (by coincidence, the deal was closed on the exact date of Maria's burial, 17 January), as well as Maria's study book for himself. Not only did Peter exhibit the paintings, but he was so impressed by them that he invited Dorothea, George and their two children to Russia. They stayed there. Dorothea designed one of the czar's major scientific exhibitions, while George became a court painter. After an initial exhibition of Maria's work, it was put away in a collection contained in the libraries of St Petersburg. Before he went on his great adventure with Vitus Bering, Georg Steller (see page 51) examined them closely.

In some ways, the locking away of the paintings in St Petersburg contributed to Maria's period of obscurity, but seen another way, it preserved her work, keeping it safe, as Maria's biographer Kim Todd writes, 'through time and revolution and fires and war'.

The Insects of Surinam, published in 1705, was a triumph and sold in three different versions, including a special deluxe edition with sixty hand-coloured transfer prints. The book finally captured all she knew about metamorphosis and the vibrant life of the insect world. Tarantulas and ants swarm over a guava tree; the white ghost moth, with its incredibly detailed wing patterns, swoops down over a caterpillar on a bush; colourful butterflies hover over an exotic pineapple, as another caterpillar carefully traverses the spiny fruit. It is life caught small and exceedingly beautiful.

Maria Sibylla Merian was progressively weakened by malaria and had a stroke that confined her to a wheelchair in 1715. She died on 13 January 1717, at the age of sixty-nine; when she was buried, four days later, her daughter Dorothea sold most of her paintings to an agent representing the Russian czar Peter the Great, who opened his country's first museum to exhibit them.

Although the great Carl Linnaeus appreciated her work, Maria's reputation faded, for a time, to that of a painter of comely flowers, but by the late nineteenth and early twentieth centuries scientists and historians had come to understand that Merian's work and life itself were a transformation. She studied insects in nature while most naturalists only observed dead specimens and understood the workings of metamorphosis while believing fervently in the design of God. She could have stayed a housewife and mother and occasional painter; instead, she became fully herself—the ultimate metamorphosis.

CARL LINNAEUS
'God created, Linnaeus arranged'

Born in Sweden in 1707, Carl Linnaeus believed that God had brought him into the world to bring order to chaos. As he was later to say, 'God created, Linnaeus arranged', a true statement that nonetheless gives you a quick snapshot of the man's outsized ego. Linnaeus was a brilliant botanist, but his main contribution—one that has outlasted the work of every single other botanist in the world—was to name things in a way that everyone could understand. He lived at a time when even simple natural things had a profusion of ungainly handles. The tomato, for instance, was called *Solanum caule inerme herbaceo, foliis pinnatis incisis, racemis simplicibus.* Quite a mouthful, and by the time you have finished saying it, the fruit will have gone bad.

Linnaeus's binomial naming system—genus first, then species—is the same one *Homo sapiens* uses today, with some modifications, and generations of tomato-eating high school students are grateful for it. His famous book *Systema Naturae*, first published in 1735 at all of fourteen pages long, went through twelve editions and ended up expanding to an incredible 2300 pages by the time Linnaeus, by then world-famous, died in 1778. Linnaeus had not only named everything, he had presented the world with the bold-faced statement that plants reproduce sexually and thus can be further classified by sexual characteristics. This caused no end of tut-tutting—one contemporary English naturalist accused Linnaeus of 'nomenclatural wantonness'—which the earthy Linnaeus simply laughed off.

However, in 1732, fame and controversy were still in the future for Carl Linnaeus. The hardy twenty-five-year-old was setting off on a solo trip through the strange northern country of Lapland— home of reindeer, shamans, savages and the midnight sun—and keeping a careful journal of this, his first great adventure into nature. The journal, finally published in 1811, provides a perfect portrait of the botanist as a young man. It is not necessarily about plants—the naturalist Richard Jefferies called it 'the best botanical book, written by the greatest of botanists ... and it contains nothing about botany'— but about the natural world and one man's place in it.

'The month of growing'

Carl Linnaeus was born to a Lutheran minister, Nils Linnaeus, and his wife, Christina, in the rural Swedish town of Rashult, in May of 1707—'between the month of growing and the month of flowering', as Linnaeus later wrote. He was a strange, fey infant—his hair was as white as snow—but curious. His father had a large garden at the back of his rectory and when Linnaeus grew into a boy he loved to spend time there among the plants. Although his father wanted him to follow his footsteps into the ministry, Linnaeus felt that his calling was in natural history. In 1727 he went to the University of Lund, in southern Sweden, to study medicine (most doctors at the time were, of necessity, botanists, for they needed to know which plants and herbs were best to alleviate which illness). The small university (which his father had attended) did not suit Linnaeus, and the next year he moved to Uppsala University, northwest of Stockholm, where he finished his medical studies and was also able to study under famous botanists such as Olof Rudbeck the Younger and Olof Celsius. The latter—whose nephew Anders would invent the centigrade temperature scale—gave Linnaeus room and board at a time when the budding botanist was too poor even to afford shoes.

With the aid of these patrons, Linnaeus became curator of the school's botanical garden and was eventually named an adjunct professor at Uppsala. He lived with Olof Rudbeck, then nearly seventy years old, and also gave private lessons to his children. Through his studies, Linnaeus had already hit upon the idea of sexual classification of plants—'In these few pages,' he wrote in the preface to his student thesis, 'I treat of the great analogy which is to be found between plants and animals, that they both make their families in the same way.' Celsius and Rudbeck were greatly impressed.

As the summer of 1732 approached, Linnaeus—casting about for a useful pursuit—conversed with Rudbeck about his lack of field experience as a naturalist. Rudbeck had taken a famous 1695 expedition to Lapland—the Arctic region in the far north of Sweden—and had written extensively about it, although all his manuscripts and notes on the trip had been destroyed in a great fire in Uppsala in 1702. He encouraged Linnaeus to retrace his steps, suggesting he apply for financial aid from the Scientific Society of Uppsala. Linnaeus did so and received a small amount of money—it would turn out to be nowhere near enough, but it was all the encouragement he needed.

'Frightful apparitions'

At the time, Lapland was a great mystery to most Swedes. Much of this rugged territory, which also extended into Norway and Finland, was unexplored. The Lapps—the Sami people—were thought of as savages, pagans who herded reindeer and lived in animal skins. There was a good deal of prejudice against the Sami. One writer who had visited them wrote: 'In Lapland, the people are dirty, flat-headed, wide-mouthed and small; they huddle around the fire frying themselves fish, croaking and shrieking.'

The Sami had a fearful reputation for magic and sorcery. Joannes W. Scheffer's 1674 work about the region contains a typical passage:

> *The melancholic constitution of the Laplanders renders them subject to frightful apparitions and dreams, which they look upon as infallible presages made to them by the Genius of what is to befall them. Thus they are frequently seen lying on the ground asleep, some singing in full voice, others howling and making a hideous noise not unlike wolves.*

The world in a name

In the seventeenth and eighteenth centuries the scientific world was involved in a long debate about how to name the plants and animals of the natural world—for progress to be made in biology, a universal framework for classification was needed, or all would be chaos. Aristotle had postulated that plants and animals should be identified using a certain set of 'essential' characters, but this was not much use to scientists in an Enlightenment world when more and more new living creatures were being discovered. In the seventeenth century, the English clergyman John Ray created systems of classification based on scientific characteristics—a step forward—but names tended to be lengthy and cumbersome.

Enter Linnaeus. In 1753 he published his Species Plantarum, *which set forth his binomial nomenclature in regard to plants: everything could be boiled down to two Latin names: the genus and the species. This was opposed to John Ray's polynomial system, which applied numerous different descriptive names to plants of the same species. A* genus, *as Linnaeus wrote, was a group of plants (or animals) sharing a unique set of common characteristics—for instance,* Homo. *A* species *denotes different living things within the genus—for instance,* Homo sapiens *as opposed to* Homo neanderthalensis. *A genus can be quite broad—*Solanum *is the Latin word for a genus that includes trees, shrubs and certain herbaceous plants. This genus contains 1400 species, of which the common potato,* Solanum tuberosum, *is one.*

Some of the Sami 'melancholy' may have come from the fact that Christian missionaries had long been in the country trying to convert them. Linnaeus noted on this journey that he saw Sami being forced by ministers to travel 80 kilometres (50 miles) to church services over rough terrain, so that they had to leave home on Friday for a Sunday service. And when a Sami would not give up his magic drum—a hollow, skin-covered piece of wood used for augury or divination purposes—he was held down and had the main artery in his arm opened with a bleeding instrument. If he did not hand over the drum, he would bleed to death. (Linnaeus notes dryly that this technique was 'often successful'.)

Aside from the Sami, however, the region abounded in natural wonders—rivers, lakes, volcanoes, plant and flower 'rarities', as Linnaeus called them—and the young man could not wait to take off. He was determined to see everything and write it down in his journal, which he hoped the Scientific Society would publish. And so, on 12 May 1732—'I was twenty-five years old, all but a half a day'—he left the city gates of Uppsala, heading north, exulting in the feel of spring, 'when Nature wore her most cheerful and delightful aspect'.

To Umea

In keeping with his determination to note down everything, Linnaeus gives us a description of himself:

> *My clothes consisted of a light coat of Westgothland linsey-woolsey cloth without folds, lined with red shalloon, having small cuffs and a collar of shag; leather-breeches [quite practical for roaming through rugged woodland]; a round wig; a green leather cap, and a pair of half boots. I carried a small leather bag ... furnished on one side with hooks and eyes so that it could be opened and shut with pleasure. This bag contained one shirt; two pair of false sleeves; two half-shirts; an inkstand, pen case, microscope and spying glass; a gauze cap to protect me from the gnats; a comb; my journal, and a parcel of paper stitched*

together for drying plants ... I wore a hanger [a short sword] and carried a small fowling piece as well as an octangular [walking] stick, graduated for the purpose of measuring.

Thus the well-equipped naturalist set forth, following a post road generally to the northeast. He was not a good horseman and could only afford a rather inferior animal, and the going was slow. Linnaeus was constantly dismounting to record points of interest, such as Runic inscriptions he saw on rocks, or to make botanical observations, some of them sharp-tongued: 'The forest abounded with the Yellow Anemone (*Anemone renunculoides*) which many people consider as differing from that genus. One would suppose they had never seen an Anemone at all.'

As Linnaeus journeyed he met his first Sami—seven reindeer herders—who spoke Swedish, unlike most of the Sami he would encounter deeper in Lapland. He also climbed Mount Nyaeckersberg, a locally famous alpine icon. The descent was so precipitous he was forced to slide on the seat of his pants for much of it (his leather breeches came in handy) but this did not stop him from plucking a young horned owl from its nest on the way down.

A few days later, on 20 May, Linnaeus climbed up to an inaccessible cave, guided by reluctant locals, and was nearly killed by 'a large mass of rock' that broke loose and landed just where he had been standing a few moments before. Despite the time of year, it began to snow that night. Finally, on 24 May, he arrived at Umea, the capital of the northern province of West Bothnia. As Linnaeus rode up, he noticed that 'in the cornfields lay hundreds of Gulls ... of a sky-blue colour'.

Umea was a sleepy town on the coast of the Gulf of Bothnia, which separates Sweden from Finland. Linnaeus rested there and met the governor, who showed him items from his cabinet of curiosities, including an otter that was so tame that he would not even eat live fish. As a man doing a great deal of hard travelling, however, Linnaeus was more interested in a type of boots, called *kangor*, worn by the locals.

These were waterproof and had no heels, which Linnaeus applauded: 'Nature, whom no artist has yet been able to excel, has not given heels to mankind, and for this reason we see the people [of Umea] trip along as easily and nimbly in these shoes as if they went barefoot.'

'In this dreary wilderness'

Linnaeus rested in Umea for two days, and then set out again. He turned away from the regular road leading around the coast and headed due west, into the primitive wilderness of the Lycksele Lapps, a land that had remained mainly unexplored. Now there were no inns at regular intervals, or places where he could trade his horse for a fresh one. He wrote, 'it became necessary for me to entreat in the most submissive manner when I stood in need of [a horse]'. To make matters worse, the weather turned bad, raining heavily, and the road became narrow and rocky. 'In this dreary wilderness,' Linnaeus confided to his journal, 'I began to feel very solitary.' It did not help, the sensitive young man wrote, that 'the few inhabitants I met had a foreign accent, and always concluded their sentences with an adjective'.

That evening he met a few Sami, who, when he asked for something to eat, gave him the breast of a woodcock that had been killed, he thought, the year before. Yet it actually tasted delicious. The rain had become so torrential that he could not continue and stayed the night at their home, sleeping beneath a very warm reindeer comforter. The next day, the rain prevented him travelling until noon, when he set off again, but soon he hit the worst road he had ever seen.

> *[There were] stones piled on stones, among large entangled roots of trees. In the interstices were deep holes filled with water by the heavy rains. The frost, which had but just left the ground, contributed to make matters worse. All the elements were against me. The branches of the trees hung down before my eyes, loaded with raindrops, in every direction. Wherever any young birches appeared, they were bent down, so that they could not be passed without the greatest difficulty.*

Still, Linnaeus persevered, and finally he made it to Granon, where he had hoped to meet some of the Sami people. He arrived for church at 9 am, but it turned out the pike were running, an occurrence that displaced even threats of eternal damnation for the Sami.

The rest of Linnaeus's journey to Lycksele Lapland would be by water, down the Umea River, and so the next day he embarked on it, with the weather finally cooperating. As he got deeper and deeper into the wilder areas of Lapland, Linnaeus noticed curious things. As his boat coursed along the Umea, he saw several owls hanging on ropes from trees; he asked his boatman why this was and was told that it was in punishment for laying their eggs in artificial nests that the Sami had created in order to capture ducks. His boatman (a Finn, not a Sami) told him that cranes were another nuisance; to warn them away, he had shot one and nailed it against the wall of his hut. 'What an absurdity', wrote Linnaeus, but the absurdities of wild Lapland were just beginning.

'A brood of frogs'

On the evening of 29 May, Linnaeus arrived in the small mission-station in Lycksele and presented himself at the home of the pastor, who with his wife suggested he stay with them until the next prayer day at church, because the Sami here were liable to 'present their firearms at any stranger who came upon them unawares' (i.e. to shoot them) and they wanted Linnaeus to be introduced to some of the locals before going out by himself. However, the next morning his hosts realised that the spring floods, abetted by heavy rains, would soon be upon them, making the road Linnaeus must take nearly impassable, so they suggested he indeed move farther into the interior. Before he did so, however, he had a chance to visit the church, which he described as being in such a 'miserable state' that if it rained the congregation 'were as wet as if they had been in the open air'. The place, Linnaeus wrote, looked like a barn. 'The seats were so narrow that those who sat on them were drawn neck and heels together.'

Here Linnaeus also met a Sami woman who claimed to have 'a brood of frogs in her stomach, owing to her having, in the course of the previous spring, drunk water which contained the spawn of these animals'. She also declared that she could hear them croaking and needed to drink a lot of alcohol to drown them out. (Linnaeus seemed to take the woman at face value and suggested she try an emetic.)

After his brief sojourn with the pastor, Linnaeus set off again, carrying with him three loaves of bread and some reindeer tongue. Travelling again by boat, he was able to take note of numerous items of natural science—the preponderance of fir and birch in the forest, the fact that the water was tinted with a reddish sediment, and the Sami's summer hunting huts, which Linnaeus sketched into his journal. Linnaeus also noted that the Sami were 'very fond of brandy, which is remarkable in all people addicted to fishing'.

The beginning of June found Linnaeus continuing his journey, travelling most of the night, 'which was as light as day' in these latitudes, 'the sun disappearing for about half an hour only'. During the long journeying, he continued to make observations in his notebook about the people he met. He described, and sketched, the conical Sami dwellings, and he wrote about the curious customs of the people who dwelled within them: 'The Laps lie stark naked, with only reindeer skin coverlets. There is no embarrassment when a man or a woman stands up naked to dress.' The people were often plagued, Linnaeus wrote, by an illness called *ullem,* a colic-like illness caused by drinking bad groundwater—the pain 'was so violent that they crawl on the ground'. Their cure, Linnaeus wrote, was to take 'soot, tobacco, [and] salt'.

Not all was squalor and pain, however—Linnaeus remarked upon a fungus (*Boletus suaveolens*) that grows on willow trees and that young Sami men placed, dried, in their pouches when they went courting—the 'grateful scent' of the fungus was apparently much desired by Sami women.

Andromeda

Leaving the river for the next stage of their journey, Linnaeus and his guide now found the going hard indeed, crossing marshland where they sank up to their knees in muck. Finally, after stumbling around for hours (it was apparent Linnaeus's guide was lost), they came upon a lonely tent inhabited by a Sami who said he would go and find a guide for them. He returned with a most unusual person:

> *He was accompanied by a human being, but whether man or woman I could not at first decide. No poet can ever have portrayed a Fury to compare to her; she might have come from the Stygian regions. She was very tiny, her face blackened by smoke. Her eyes were brown and sparkling, her eyebrows black; and her jet-black hair hung loose about her head, on which she wore a flat red cap. Her dress was grey; and from her chest, which was like the skin of a frog, were suspended a pair of long limp brown dugs.*

This 'Lapp Fury', as Linnaeus went on to call her, was actually quite compassionate. She could not understand why anyone would come voluntarily into her country ('O you poor man! What cruel fate has brought you here? Wretched man!'). She told him that ahead was nothing but a flooded river and there was nowhere that he might find anything to eat. Linnaeus, already sick from drinking polluted water and weak from hunger, saw that there was no more point in going on, and, along with his boatman, turned back. Even this was fraught with peril—the river was so high with floodwaters that their boat was stove in by a rock and they were forced to walk the rest of the way back to Lycksele, where Linnaeus rested with the pastor and his wife, and then returned to Umea. After four days there, Linnaeus, undaunted, set out again. This time he followed the coastal road that he had left before his side trip to Lycksele. His goal was to journey to the small town of Lulea, and then head west again into the mountains of Lapland, which he wanted to climb to see the sun above the horizon at midnight.

Along the coast north of Umea, the young naturalist experienced both the beauty and cruelty of Lapland. He came upon wild fields of a shrub with beautiful, pink flowers that he named 'Andromeda', after the princess from Greek mythology. 'I noticed that she was blood-red before flowering, but that as soon as she blooms her petals become flesh-coloured', Linnaeus wrote. 'I doubt whether any artist could rival these charms in a portrait of a young girl, or adorn her cheeks with such beauties as are here and to which no cosmetics have lent their aid.' As if to prove his point, Linnaeus illustrated his notebook with a picture of the andromeda (a plant also known as the bog rosemary, which is quite common in the British Isles), alongside a nubile young maiden.

However, about 250 kilometres (150 miles) farther up the road, Linnaeus came upon 'a gibbet with a couple of wheels upon which lay the bodies of two decapitated Finns who had been executed for highway robbery and murder. Beside them was the quartered body of a Lapp who had murdered one of his relations.' Linnaeus had probably seen similar sights in Uppsala—it was the harsh way crime was dealt with at the time—but the sight was disturbing and he hurried quickly by. Yet even these mutilated bodies could not turn him away from the beauties of the northern summer, for on 24 June he wrote: 'Midsummer day. Blessed be the Lord for the beauty of summer and of spring and for what is here in greater perfection than almost anywhere else in the world—the air, the water, the verdure of the herbage, and the song of birds!'

Sixteen naked Lapps

The day after this paean to summer, Linnaeus turned westwards again, heading once more deep into Lapland. This journey was, at first, notably more comfortable. He travelled by boat up the Lule River and into Lake Skalka, where he reached the town of Kvikkjokk. The pastor's wife gave him foodstuffs for a week, and two Sami guides and, on 6 July, Linnaeus set off to ascend Mount Vallevare.

On the way, he continued to write about the Lapp reindeer, whose habits—milk production, skins, herding—he found quite fascinating. He had noticed that their hoofs made a strange clacking sound when they were herded over snow—it sounded like Spanish castanets—and he now discovered that their hoofs were hollow. The Sami were no less objects of his attention—they rubbed dog's fat onto their sore backs (preferably 'while sitting before a fire') and used no razors but cut their beard with scissors. They never cut the hair on their heads. They smoked a great deal, and he found he could always trade a small amount of tobacco for a big round of cheese. And they did not kiss in greeting, instead rubbing noses (actually, the Sami rub cheeks).

After climbing the 2500-metre (8000-foot) high Mount Vallevare, Linnaeus had a view of snow-covered mountains: 'Nothing but mountains, each one bigger than the last, and all covered with frozen snow. There was no road, no track, no sign of human habitation. Summer's green seemed to be vanished ...' Coming back down, he exulted:

> *I scarcely knew whether I was in Asia or Africa ... the soil situation and every one of the plants being equally strange to me ... All the rare plants I had seen before and rejoiced in were here as if in miniature; indeed there were so many that I feared that I was taking away more than I would be able to deal with.*

Now completely obsessed with collecting and sorting these plants, Linnaeus let time slip by until one of his Sami guides warned him that they still had 65 kilometres (40 miles) to go to 'the next Lapp tent', where they were supposed to shelter. There followed a tense and surreal scramble through this mountainous winter landscape, all of it under the pallid rays of the midnight sun. Finally, on the evening of the next day, they arrived at the tent, which was a round yurt-like hut, covered with skins, with more reindeer skins on the floor. His guide led him inside and Linnaeus described the scene in his journal:

The inhabitants, sixteen in number, lay there all naked. They washed themselves by rubbing their body downwards, not upwards. They washed their dishes with their fingers, squirting water out of their mouths upon the spoon, and then poured into them boiled reindeer's milk, which was as thick as common milk mixed with eggs, and had a strong flavour ... My hosts gave me missen to eat; that is, whey, where the curd is separated from it, coagulated by boiling, which renders it very firm. Its flavour was good, but the washing of the spoon took away my appetite, as the master of the house wiped it dry with his fingers, whilst his wife cleaned the bowl, in which milk had been, in a similar manner, licking her finger with every stroke.

These Sami, while not up to Linnaeus's sanitary standards, were quite friendly and lived by herding reindeer. Linnaeus grew to be fond of them, and to admire their healthiness—for health was a subject that obsessed the great naturalist all his life. He noted with some chagrin that his Sami guides were men in their sixties and seventies who did not get tired out during their long journeys, whereas he—a young man in the prime of his health—was often too exhausted to take another step. He felt that their physical activity—'I never saw one with a big belly'—combined with a steady diet of milk and meat kept them fit.

'After all I have suffered!'

Linnaeus journeyed onward, walking west through the mountains (at one point falling into a crevasse and having to be pulled up by ropes) and into Norwegian Lapland and to the coast, where he rested for a time with a Norwegian pastor and his beautiful daughter, Sarah, with whom the young naturalist was quite taken. (Sarah, for her part, told him that 'she never expected to meet on honest Swede' but here, at last, was one. The smitten Linnaeus took this as quite a compliment.)

On 15 July, Linnaeus at last began his journey home, retracing his steps over the mountains and arriving back at Kvikkjokk on Lake Skalka. There were no boats to take him down the River Lule and so he and a guide fashioned a raft, which unfortunately sank within a few kilometres. They were able to scramble ashore, but only after Linnaeus lost several specimens from his collection.

Arriving at the shores of the Gulf of Bothnia, Linnaeus went north to the town of Tornio (in Finland) because he had decided he wanted to take passage by sea back to Stockholm and then to Uppsala. But with vessels being delayed by contrary winds and the season turning late for journeying on the often storm-tossed gulf, Linnaeus decided to return along the east, or Finnish, coast. It was not a part of his adventure that he especially liked—there was little new to botanise and almost no one spoke Swedish. He felt the Finns were lazy and slovenly and disliked their low, smoky huts. ('If *I* had management of these Finns,' he scribbled in his journal, 'I would tie them up against a wall and flog them until they promised to build chimneys.')

Finally, Linnaeus reached Turku, Finland's coastal capital at the time, and found passage across the Gulf of Bothnia, arriving in Uppsala at last on 10 October, after having been away nearly five months. He had covered nearly 5000 kilometres (over 3000 miles) by foot, horseback and boat, and his expenses had far exceeded the amount given him by the Scientific Society. He appealed to them for more money, but the cash-strapped society could only give him a small portion of what he owed. 'After all I have suffered!' he exclaimed indignantly in his journal.

However, there were consolations. The society published in its distinguished journal a portion of his *Flora Lapponica*, a compendium of the flowers and plants he had collected and categorised on his journey, and the first printed work to feature his sexual system of classification. (Linnaeus's journal itself would have to wait until 1811, some thirty years after his death, to find a publisher, and it was then printed in England under the title *A Tour in Lapland*.) Although

Linnaeus travelled far in the world of scientific thought, he spent most of his life in Sweden, a renowned professor, married with five children. His Lapland journey—which he loved to recount, greatly exaggerated, later in life—was in some ways his greatest adventure.

'Yes, Love comes even to plants'

Those who studied nature closely had long known that plants had what amounted to sex organs—the ancient Greeks, for instance, had observed that each date palm is either male or female. And the German naturalist Rudolf Camerarius had, in 1694, experimented with the reproductive organs of plants and understood that pollen was the sperm of the botanical world and that female plants not located near male plants produced fruit, but no seeds.

But no one had gone as far as the young Carl Linnaeus. In the rhapsodic beginning of his thesis for Uppsala University, he wrote:

> Words cannot express the joy that the sun brings to all living things. Now the black-cock and the capercailzie [a wood grouse] begin to frolic, the fish to sport. Every animal feels the sexual urge. Yes, Love comes even to plants. Males and females, even hermaphrodites, hold their nuptials (which is the subject I now propose to discuss) ...

In language that sounded nearly pornographic (Linnaeus described the flowers' leaves as 'the bridal bed which the great Creator has so gloriously prepared, adorned with such precious bed-curtains, and perfumed with so many sweet scents ...'), the young professor described stamens and pistils, pollination and seeds. It is no wonder that, when Linnaeus's theories about the reproductive organs of plants became well known, he incurred the wrath of those who felt threatened by such open and joyous treatment of sexuality, even in plants.

One Swedish minister, observing 'boys and girls botanizing together' was horrified by this innocent activity and demanded they stop. And the German writer Johann Siegesbeck wrote haughtily: 'Who would have thought that bluebells, lilies and onions could be up to such immorality?'

GEORG STELLER
Sunday's child

*The day was spent in tacking in order to get close to the island, to
enter the large bay seen from the distance, and at the same time to
come under the lee of the land. This was also accomplished, with the
greatest apprehension, when on Monday the 20th we came to anchor
among numerous islands. The outermost of these had to be named
Cape St. Elias, because we dropped our anchor under the lee of it on
St. Elias' day. For the officers were determined to have a cape on their
chart notwithstanding the fact that it was plainly represented to them
that an island cannot be called a cape ... Only on one point were [we]
all unanimous, viz. that we should take fresh water on board, so that I
could not help saying that we had come only for the purpose of bringing
American water to Asia.*
Georg Steller's *Journal*

One understands how Vitus Bering and his lieutenants must have
felt. Here they were, aboard the imperial ship *St Peter*, engaged in
making history—for in July of 1751 no one sailing from Russia had
ever reached the northern shores of Bolshaya Zemlya, 'the Big
Land', as America was called.

Bering was about to make landfall on a wooded, mountainous island, opening up a potentially lucrative trading route and vast territorial gains for Russia. And yet just at this historic moment this oddly dressed German pipsqueak of a naturalist was butting into conferences and telling Bering and his men (seamen with decades of experience between them) what was a cape and what was not a cape and making sarcastic comments about bringing fresh water on board.

The man just would not shut up. Finally, to quiet him down and get rid of him, Bering acceded to the naturalist's request and sent him ashore with the water party, but he could not help playing a trick on him. 'On my departure,' wrote the naturalist, 'the Captain Commander made a test as to how far I could distinguish between mockery and earnest by causing the trumpets to be sounded after me, at which, without hesitation, I accepted the affair in the spirit in which it was ordered.' But, the prickly naturalist added, as usual aiming for the last word, 'I have never been a braggart, nor would I care for such attentions even if they had really intended to honor me.'

Of course, one gets the feeling that Georg Steller—for that is the name of the naturalist—would have loved a few trumpets sounded in earnest for him, but this never happened during his short and tempestuous lifetime. Posterity, however, has recognised him as a pioneering scientist—the first trained naturalist to explore the North Pacific, the only scientist to see a live sea cow to that time, and the man who identified numerous new plants and animals, and linked life in Kamchatka with that in Alaska. Beyond that, wearing his other hat as physician, he was to save the lives of many of the *St Peter*'s crew on this American voyage by treating them with native remedies for scurvy, for Steller was the rare scientist of his day who paid close and respectful attention to how native peoples thrived in the wild.

Unfortunately, his remedies could not save the life of the great explorer Vitus Bering, but the relationship between the two men—leader and scientist—is one of the most fascinating in the history of the natural sciences.

Sunday's child

It is typical of Georg Steller that he was born under the twin moons of fortune and misfortune. He came into the world on Sunday, 10 March 1709, the second child of the Stöller family of Windsheim, Germany. Being born on a Sunday was a lucky thing in German folklore—the baby was a *Sonntagskind*, or Sunday's child, who would be blessed all his life.

There was only one problem—the infant was dead. After the midwife failed to revive him she gave up and left, but the baby's aunt wrapped up the cold child in hot blankets until, lo and behold, he gave out a startled yell. The delighted family named him Georg Wilhelm and he grew into an intelligent and headstrong boy. His family were Pietist Lutherans and raised him with a strong social conscience, a sense that if one saw wrong being done—particularly to the poor or defenceless—one had an obligation to right it.

After graduating from secondary school, where his fascination with the natural sciences had been evident, Stöller studied theology, medicine and botany at universities in Wittenburg and other German cities. He passed his examinations with honours in 1734, but there was a shortage of teaching positions in German universities—the natural path someone like Stöller might take—and, in any event, the young man wanted more adventure. He had heard of the natural mysteries uncovered by the first Siberian expedition of the explorer Vitus Bering, which had finished in 1730, and had heard, as well, that Bering was even now planning a second expedition to Kamchatka on the far eastern coast of Russia. Desperate to go along, Stöller signed up as physician to a group of wounded Russian army soldiers being transported back home from a war in Poland. Arriving in St Petersburg, he was able to wangle a job as personal physician to the archbishop of St Petersburg. There he changed his name to Steller— easier for Russians to spell and pronounce—and to the archbishop's amusement spent much of his time wandering the countryside outside the city in search of rare plants. However, through the good graces of his mentor, he was named to the Russian Academy of Sciences, which appointed him, in 1737, as an adjunct professor of natural sciences to accompany Bering's second Kamchatka expedition.

Wildly excited both with the appointment and with his salary of 630 roubles a year, Steller proposed marriage to the young and lovely Brigitta Messerschmidt, widow of a naturalist who had accompanied Bering on his first expedition. She promised to go along with him to Kamchatka but, in the end, decided to stay home—and demanded half his salary, to boot. The heartbroken Steller set out alone for the wilds of the east in May of 1738.

'I have entirely forgotten her'

Georg Wilhelm Steller was joining a great Russian drama in its second act. Vitus Bering, a Danish explorer in the employ of the Russian court, had been sent by the ailing Czar Peter the Great in 1725 on an expedition to the east, with the intention of exploring the almost unknown areas of Siberia, the shores of the Icy Sea (the Arctic Ocean) and the Kamchatka Peninsula. The czar also wanted Bering to sail east across the northern Pacific to find Bolshaya Zemlya, which most educated Russians assumed to be America. If Bering could find it, Peter wanted Russia to open the Big Land up to trade, to give Russia the same kind of stake in the New World as France, England or Spain.

Bering's first expedition consisted of thousands of soldiers, porters, Cossacks and scientists, who followed him overland to Siberia. After that Bering made his way to Kamchatka, where he constructed a ship and sailed it out into what would become the Bering Sea. But by then it was August of 1728 and the short northern summer was coming to an end. Without sighting North America, he turned back.

Bering returned to the Russian court in 1730 to find that both Peter the Great and his successor, Catherine I, had died. The new empress was Anna Ivanovna, Peter the Great's niece, who insisted that he embark on a second expedition with several daunting goals. He was to further chart the interior of Siberia and Kamchatka and settle it, setting up iron foundries, shipyards, cattle farming and even a postal system. He was also to send a ship to open up trade relations with the fabled country of Japan. And, last but certainly not least, he was to make a successful attempt to find America and chart its coastline.

It was this massive second expedition that Steller joined in 1738. Thousands of Russian settlers had trailed after Bering as he headed east, arriving in Kamchatka in 1739. He sent one of his sea captains to explore Japan while he himself made preparations to sail for America. Several prominent Russian naturalists and doctors had accompanied the expedition, but Bering's refusal to pay their expenses alienated them and they soon decided to head back to St Petersburg. In need of a botanist and ship's doctor, Bering turned to Steller.

Steller had been having the time of his life on this expedition, so much so that he wrote about his wife in his journal: 'I have entirely forgotten her and fallen in love with Nature.' After crossing the Ural Mountains in the summer of 1738, he made his way through the Siberian wilds, cataloguing all types of flora and fauna, and preserving them between leaves of paper. (Steller also pressed insects between thin sheets of clear mica, a process that kept them wonderfully preserved; scholars could examine them in exact detail many years later.)

However, and importantly, Steller was also an observer of humanity, with some training in ethnography. When he arrived at last in the Kamchatka Peninsula in October of 1740, he made a careful study of the Itelmen people who were native to the region—people whom Vitus Bering scornfully dismissed as 'idolators and ... strangers to all good customs'. It was true that some of their customs were rather curious to the eyes of Russians and Europeans—for instance, the Itelmen were highly promiscuous. Steller wrote in his journal (with some circumspection) that 'Whoever comes to Kamchatka and does not acquire a woman ... is forced to do so from necessity. No one washes or sews for him, takes care of him or does him any favors unless he pays with sexual intercourse.'

Whatever Steller paid his housekeeper with, he also learned much else about the Itelmen—how they wove grass into clothing (even stockings and raincoats), how they pioneered the first use of sunglasses (eye-protectors made of nets of birchbark and horsehide, to prevent snow blindness) and what herbs they ate to protect against scurvy in a land with few natural fruits and vegetables. Many of the Russians laughed at Steller's close study of this despised native people, but it would save their lives later on.

Steller's scurvy cure

Georg Steller was one of numerous scientists of the eighteenth century who came, independently, to the conclusion that certain types of plants, fruit and vegetables (as well as fresh, rather than salted, meat and fish) contained properties that helped keep scurvy, the dreaded disease then facing mariners, at bay.

In Steller's case, this came about during his time among the Itelmen people of Kamchatka, where he noticed no incidence of scurvy at all, despite the fact that the people spent a good deal of time above the Arctic Circle. By carefully interviewing the female shamans of the tribes, Steller was able to understand which plants had anti-scorbutic properties, as he wrote in his journal:

> Scurvy actually bothers only the new arrivals on Kamchatka, but the Cossack children and the Itelmen not at all because of their mixed diet of many roots, plants and tree barks, including wild garlic, called *cheremsha*, the yellow and black scurvy leaves, called *moroshki*, and *shiksha* [chokeberry], and frozen fish eaten raw ... A decoction of shrub pine, called *slanets*, is very useful and obviously effective; also a decoction made from the buds of the stone alder shrubs has an even greater effect and a very pleasant aroma ...

When Steller first gathered such plants during the St Peter's *stop on Nagai Island, the crew refused to take them; it was only after a number of men had died that they would allow Steller to minister to them.*

Voyage to North America

Bering's interview with Steller went well and he hired the young naturalist on the spot, with a haste that he may later have regretted. When Bering's two ships, the *St Peter* and the *St Paul*, set sail to find North America in June 1741, Steller accompanied the Danish captain on the *St Peter*. He soon made it known that he was no meek and retiring physician and scientist. First of all, he told Bering that Bolshaya Zemlya lay to the northeast (as it did). He based this on conversations with the Itelmen, who told him that large trees drifted across the ocean from this area and that land could be seen on exceptionally clear days. But, after scornfully replying, 'People talk a lot. Who believes Cossacks?', Bering headed his little flotilla southeast. After eight days at sea and Steller's constant badgering—Steller wrote that the officers of the *St Peter* replied to his advice by saying 'You do not understand ... you are not a seaman'—Bering finally turned to the northeast. No sooner had he done this, however, than a storm arose and separated him from the *St Paul*, which would eventually return safely to Kamchatka.

The *St Peter* journeyed on, stuck in heavy fog, until, as Steller wrote, 'We saw land as early as July 15, but because I was the first to announce it and because [the land immediately disappeared in the mist], the announcement, as usual, was regarded as one of my peculiarities.' However, the next day the sun broke through and a snow-capped mountain loomed 5500 metres (18,000 feet) high on the horizon. They were off the coast of present-day Alaska; the mountain was Mount St Elias. A few days later, Bering tacked until they reached the shores of what is now Kayak Island and, after sarcastically blowing his trumpets, allowed Steller to head ashore for a few hours while water barrels were being filled.

It was a triumph for the natural scientist that Steller was able to spend even the few hours he did on the North American continent. He immediately found the warm remains of native fires—for the locals had apparently fled at the sight of the Russians—and traced similarities between what he located and the customs of the Itelmen.

These North Americans 'cooked their meat by means of red-hot stones', as did the people of Kamchatka. Both peoples also ate large amounts of fish (used in lieu of bread) and blue mussels, and made their fires by means of wooden, friction-creating fire drills.

Steller's observations of the natural world were just as valuable. He found a 'new and elsewhere unknown species of raspberry', which, while still green, looked to be quite juicy and delicious. 'Of birds,' Steller wrote, 'I saw only two familiar species, the raven and the magpie; however, of strange and unknown ones I noted more than ten different kinds.' These included a bright blue bird that his huntsman was able to shoot 'of which I remembered to have seen a likeness, painted in lively colors and described in the newest account of birds and plants of the Carolinas [the work of the English nature writer Mark Catesby]'. This bird is now named after Steller—Steller's jay—and it was this sighting that 'completely proved to me that we were in America' and provided him with the broad picture of a blue jay travelling across the span of a continent.

Finally, as the day drew to a close, Steller sent an urgent message back to Bering that he needed more time to explore the wonders of this island. Steller wrote down that Bering's 'patriotic and gracious reply' was to 'get his butt back on board' the *St Peter*, which Steller resentfully did. However, once on deck he was greeted with a rare cup of hot chocolate, a delicacy saved for important occasions. It turned out that Bering had lost sight of Steller as he wandered the island and had been somewhat concerned that ill fortune might have befallen him. Steller irritated him, but Bering could not help but admire the man's single-minded tenacity. Besides, one gets the feeling that Steller sort of grew on one the longer one spent in his company.

Fateful journey

Bering then headed the *St Peter* back towards Russia, but instead of taking a direct route sailed southwest, roughly following the course of the Aleutian islands, which trail in a half-moon out from the coast of North America. This was not entirely Bering's idea—his officers (who, under Russian law, had a good deal of power to make decisions on board) insisted that his charter with the czarina demanded that he explore these islands. But contrary winds and the necessity of stopping frequently slowed the *St Peter* down a good deal.

Even worse, scurvy now rode aboard the vessel. One by one, seamen experienced the debilitating weakness and loosening of the teeth that came with the disease—Bering himself was suffering from it. When the ship stopped at bleak Nagai Island to take on water, Steller recorded how the seamen had chosen 'the first and nearest stagnant puddle and already started operations'. Knowing that the water was 'stagnant and alkaline' and that it would dehydrate the crew further, thus making them more vulnerable to disease, Steller found a source of purer water farther inland, but he could not convince the crew that it was worth the effort to drag their heavy barrels far enough to collect it. Giving up, Steller reconnoitred the island and found numerous of the anti-scorbutic herbs he had learned to identify while in Kamchatka, but he could not find one sailor who would help him pick them in the quantities they needed: 'I had ... requested the detail of several men for the purpose of collecting such quantity of anti-scorbutic herbs as would be enough for all, nevertheless, even this proposition, so valuable to all and for which I merited gratitude besides, was spurned.' He added ominously: 'Later, however, there were regrets enough, when ... I was tearfully begged to help and assist.' Steller took only enough herbs for himself, for his assistant and for Bering.

While they were at Nagai, a sailor named Shumagin died of scurvy and thus became the first European to be buried on northwestern American soil. By the time the expedition left Nagai—having finally been able to trade with Aleut Islanders, who were seeing Europeans for the first time—it was nearly September and the Bering Sea was roiled by fierce storms, where white walls of water crashed down across the ship and the phenomenon called St Elmo's fire danced in the mast tops. Steller's herbs had now run out and Vitus Bering was bedridden, 'without the use of his limbs', as Steller wrote. The second-in-command, First Mate Sven Waxell, fought mightily just to keep the ship afloat. Finally, in November, with the rigging of the vessel nearly gone, he brought the ship to landfall at the bleak Commander Islands, more than 160 kilometres (100 miles) off the coast of Kamchatka. When they realised they were indeed on a large island and not on the mainland itself, the exhausted and sick officers and crew settled in to weather what would be a long and terrible winter.

The Island of the Foxes

The Commander Islands are a group of seventeen barren islands, and the island on which Steller and his companions found themselves was the largest—90 kilometres (56 miles) long and 24 (15) wide. The Commanders were uninhabited and their desolate landscape consisted mainly of sand bluffs and scrub brush. While the crew would later rename this large island Bering Island, for the time being they referred to it bitterly as the Island of the Foxes. As the sick and dying crew dug makeshift homes—holes in the side of a sand bluff covered with sailcloth—they were plagued by arctic foxes that stole their food and even tried to eat the dead. Steller records:

> *The dead, before they could be interred, were mutilated by the foxes, who even dared to attack the living and helpless sick, who lay about the beach without cover, and sniffed at them like dogs ... The blue foxes, which by now had gathered about us in countless numbers, became, contrary to habit and nature, at the sight of man more and more tame, mischievous, and to such a degree malicious that they pulled all the baggage about, scattered the provisions, stole and carried off from one his boots, from another his socks, trousers, glove, coats etc. ... It seemed that the more of them we killed and tortured for revenge most cruelly before the eyes of the others, letting them run away half-skinned, without eyes, ears, tails, half-roasted, etc., the more malevolent and audacious become the others.*

The horrors of this winter continued when the *St Peter* was attacked by a storm in the bay and essentially destroyed, making the men, or so they thought, captives on this horrible island. And here is where Georg Steller proved his mettle. Despite his quarrels with Bering and the crew, despite his own disdain for the ignorance that had caused them to ignore his advice, he became a ministering angel to the men. Along with his servant, Thomas Lepekhin, and his artist, Friedrich Plenisner, Steller worked ceaselessly to build a dugout infirmary for the sick, to bring them the anti-scorbutic herbs they needed, to shoot game for them, and to keep them as warm and dry as humanly possible. His special patient was Vitus Bering. The sixty-year-old captain-commander was now quite ill, despite Steller's efforts to feed him special food. One day in November, Steller wrote: 'Today I brought the Captain Commander a young, still suckling sea otter and counseled him in every way and manner to let it be prepared for himself in default of other fresh food, but he showed a very great disgust at it, and wondered at my taste ...'

Bering's refusal to eat fresh, raw meat probably helped doom him, although the observant Steller stated that as Bering lay dying he still had all his teeth, which may rule out scurvy as the actual cause of the great captain-commander's death. Bering just seemed to sink, covered in warm sand, staring out at the rolling sea with what Steller called in amazement 'composure and strange contentment'. He died on 8 December, possibly from scurvy but also, as Steller wrote, 'from hunger, cold, thirst, vermin and grief'. One of the last things he said to Steller was: 'The important thing now is to take care of the men', and this is what Steller vowed to do.

'Our health noticeably improved'

From December 1741 to May 1742, Georg Steller laboured to take care of the crewmen Bering had left behind. They, for their part, began to appreciate him. Sven Waxell, now in charge of the expedition, wrote in his own journal:

As soon as the snow was gone, and green shoots came out of the ground, we collected and used quantities of herbs and plants. In this, Adjunct Steller gave excellent assistance, for he was a good botanist. He collected and showed for us many green herbs, some for drinking, some for eating, and by taking them we found our health noticeably improved.

The much-maligned Steller had, at last, proven his worth to these tough seamen. And, while he was engaged in nursing the crew, Steller did not neglect his duties as a naturalist. With astonishing energy, given the fact that he was half-starved himself, he explored the Island of the Foxes which, due to its isolation, had long acted as a kind of nature preserve for numerous species. He risked his life climbing high cliffs to study the creature that became known as Steller's sea eagle ('While I was inspecting a nest ... the two parents attacked me so forcibly I could barely fend them off with a stick'). He also became the only scientist to describe the flightless spectacled cormorant, which was as large as a goose and 'from the ring around the eyes, and the clown-like twisting of the neck and head, it appears quite a ludicrous bird'.

This creature would be extinct within thirty years, as would the next animal Steller described—known as Steller's sea cow. Steller was the only scientist ever to study this animal. This extraordinary creature was a northern manatee, averaging about 8 metres (26 feet) in length from its tiny head to its whale-like tail and weighing, in full adulthood, up to 8 tonnes. The animal had no teeth, per se—only what Steller likened to 'two white bones' that were used for grinding the kelp it ate. It spent a great deal of its time half-in and half-out of the water, lying near freshwater streams that poured into the ocean, eating plant life. Steller constructed a blind and watched the animals, which travelled in herds, closely:

These gluttonous animals eat incessantly ... All they do while feeding is to lift the nostrils every four or five minutes out of the water, blowing out air and a little water with a noise like that of a horse snorting.

While browsing they move slowly forward, one foot after the other, and in this manner half swim, half walk like cattle or sheep grazing. Gulls are in the habit of sitting on the backs of the feeding animals feasting on vermin infecting their skin ... The animal has no hair ... It has bristles rather, or hollow quills, and these are found only around the mouth and under the feet ... The feet are entirely without claws ... so that the animal moves on a skin that is rough with bristles.

Unfortunately for Steller's sea cow, its flesh was delicious, tender and veal-like; its fat was 'as pleasantly yellow as the best Dutch butter'. One of the creatures would feed the entire ship's company for a week. It was the same reason aboriginal hunters would soon hunt the sea cow out of existence—a process that had been going on before Steller arrived on Bering Island.

'As in a dream'

Beginning in April, the ship's crew, led by Sven Waxell, began to build a new ship, using what timber they could salvage from the wreck of the old *St Peter* and caulking its seams with sea-cow fat. By 13 August the work was done. Before boarding the ship, the men built a wooden cross to honour Vitus Bering and named the island Bering Island (and the island group the Commander Islands). They set sail, arriving in Kamchatka's Avacha Bay, from which they had set sail, on 27 August. Steller wrote in his journal of the mixed blessing of homecoming:

Great was the joy of everybody over our deliverance and safe arrival ... We had been regarded by everyone as dead or lost; the property which we had left behind had fallen into the hands of strangers and had mostly been carried away. Therefore, in a few seconds, joy turned to anxiety in the hearts of all of us. However, we were all by this time so much used to misery and sorrow that, instead of looking forward anew, we only thought of continuing the old life and regarded the present circumstances as in a dream.

While the crew dispersed, the indefatigable Steller spent the next year working feverishly, preparing and annotating his journals and travelling restlessly through Kamchatka. He journeyed near the Arctic Circle to study the Olyutor Koriaks, a tribe known for its fierceness (they had a habit of killing such pests as Russian tax collectors). However, they allowed Steller to live among them and in the spring join their whale hunts. Returning south by dogsled that spring, he noticed that the sea was still frozen all the way out to Karaga Island, which had never been visited by a naturalist. He decided to take his sled across the ice but, weakened by the warming of the weather, the ice broke when he was only halfway across. Here, once again, his study of native peoples stood him in good stead—he was able to don snowshoes and hop back to shore on ice floes, just as he had seen hunters do in the Kurile Islands.

In the spring of 1744, Steller had returned to live among the Itelmen, whom he loved, and was alarmed to find that they had been subjected to brutal repression by the Russians and the Cossacks who acted as Russian soldiers and police. Steller protested at the imprisonment of four of these Itelmen and took it upon himself to let them go home to their villages. Then he began to travel back to Russia but was detained by a vengeful Russian officer and charged with treason. By the time his name was cleared, Steller, whose health had perhaps finally been weakened by so many years of hard work and poor diet, had begun to drink heavily. In the fall of 1746, he caught a fever near the town of Tyumen and died there on 12 November, at the age of thirty-seven.

Like Vitus Bering, Georg Steller was not immediately appreciated for the important body of work he left behind, but his pioneering skills in understanding and empathising with native peoples and his contribution to natural science is unchallenged. And anyone who reads his acerbic and often quite funny journals will appreciate the picture of a brilliant and impatient man, surrounded by those whom he considers dullards, who is just trying to get the job done properly.

Bering: the forgotten explorer

Despite the fact that the Bering Strait, which separates the northernmost reaches of Asia and America, remains named after him, Vitus Bering is not an explorer ever named in the same breath as other seafaring adventurers like Ferdinand Magellan and James Cook.

Part of this may be because he explored the frozen North, but a great deal of it is because he was a man without a country. Vitus Bering was born in a small port town in Denmark in 1681. At fifteen, he shipped out as a cabin boy and spent almost a decade sailing the world with the Danish merchant fleet. While taking officer training, he came to the attention of a Russian admiral sent to Scandinavia by Peter the Great to search for qualified naval officers—it was Peter's dream to take Russia out of its landlocked status by building a colossal fleet, and to do this he needed qualified foreign officers to lead and to train Russians.

Hired by the Russians, in 1724 the czar offered Bering the opportunity to explore Siberia, Kamchatka and North America. But after his first expedition ended in 'failure' (because he had not reached America), Bering found out how capricious the Russian court could be. He was forced back to the east on an almost impossible mission to settle the areas he had previously explored and to find 'the Big Land'. He knew that to return without achieving the latter would mean the end of his career and possibly even imprisonment.

Ironically, he did find North America but died on his way home. Because Russia was experiencing a wave of xenophobia, the new ruler, Empress Elizabeth, decided to keep secret the fact that Bering had even found America. Even when it was accepted that he had indeed landed on Kayak Island, he was portrayed as an over-cautious functionary without a true explorer's spirit. It was only in following centuries that Bering's true achievements in mapping the northern lands of the Pacific and opening up Siberia and North America to Russian traders (eventually leading to the purchase of Alaska by the United States in 1867) were recognised.

In 1992, in a ceremony attended by both Danes and Russians, a statue was erected to Vitus Bering on lonely Bering Island, facing out across the strait that bears his name.

WILLIAM BARTRAM
'Plant Hunter'

It is a conflict often seen between accomplished fathers and their bright but unfocused sons. Young Billy Bartram showed talent as an artist and was fascinated by plants but, thought his father, the boy was too unrealistic to support himself. 'It is now time to prepare some way for [Billy] to get his living by,' he wrote to a friend in 1755. 'I want to put him to some business by which he may, with care and industry, get a temperate, reasonable living. I'm afraid that botany and drawing will not afford him one, and hard labour don't agree with him.'

And John Bartram's friend wrote back: 'I am concerned that Billy— so ingenious a lad—is, as it were, lost in indolence and obscurity.'

The object of this attention was sixteen years old at the time, a student at the Philadelphia Academy (later to become the College of Philadelphia). Because his father was well known, Billy had any number of opportunities that other young men would have killed for. Benjamin Franklin offered to take him into the printing business; his father told him he would finance Billy's training as a doctor; and Peter Collinson, the English horticulturalist and Fellow of the Royal Society who had worried about Billy's 'indolence', suggested he work as an engraver, an occupation that would allow him to keep drawing.

Stubbornly, Billy Bartram said no to all these things, half-heartedly tried his hand in the business world, and then fled the pressures of Philadelphia to work for his Uncle William, who had a trading post in North Carolina. That did not work out and suddenly it was 1765, Billy was twenty-five years old and had not a prospect in sight. 'Nothing but marrying will settle him', intoned Collinson to John Bartram, who was tearing out his hair over a son whose only desire in life was to wander through the wilderness finding plants and animals to paint.

Funny, how blind fathers and sons are to each other. John Bartram was, at the time, the most famous naturalist in America and had been wandering through the wilderness ever since Billy could remember. In fact, he had taken Billy on a collecting expedition to the Catskill Mountains in New York State when the boy was only fourteen (calling him, with delight, 'my little botanist'). John was the proprietor of the finest botanical garden in America and was about to be named Royal Botanist for the colonies by King George III.

How could Billy Bartram turn out to be anything but a chip off the old block?

'Poor Billy Bartram'

William Bartram was born, seventh of eleven brothers and sisters, and twin to Elizabeth, on 20 April 1739. He grew up in a home filled with books and artwork, where his father corresponded with and was visited by some of the great thinkers of his time—not just Benjamin Franklin, but also men like Thomas Nuttal, André Michaux and Benjamin Barton. It was on the Catskill trip with his father that William's skills in drawing and watercolours first revealed themselves, delighting both John Bartram and his English patron, Peter Collinson, who showed the work to his influential friends in London. Yet William appeared to lack the drive and focus of his father, who, aside from botanising, ran a farm and supported a wife and eleven children.

When John Bartram was named Royal Botanist in 1765 he was already sixty-six years old and decided to take advantage of the stipend the job awarded him to go on one last botanising trip. This time his destination would be Florida—then an unknown wilderness to most Americans. Stopping in North Carolina at his brother's trading post, he visited young William, who begged his father to take him along. John Bartram acquiesced and the two journeyed through the wilds of Georgia and Florida—home to panthers, alligators, crocodiles, rattlesnakes, and Cherokee, Seminole, Creek and Choctaw Indians, many of whom had only recently been at war with the British crown.

Father and son spent a year on their journey, arriving in the old Spanish city of St Augustine, Florida, and then making a 650-kilometre (400-mile) canoe trip down the St John's River, charting the course of the river for the first time and finding such plants as the Venus flytrap and the aptly named pitcher plant. ('What a quantity of water a leaf is capable of holding,' William Bartram wrote in his journal, '[and] how cool and animating—limpid as the morning dew.')

When the trip was over, John Bartram returned to Philadelphia with 'a fine collection of strange Florida plants', while William stayed on to start an indigo plantation. His father had his doubts. 'I have left my son Billy in Florida,' he wrote soon after returning to Philadelphia. 'Nothing will do with him now but he will be a planter upon the St John's River ... This frolic of his hath drove me to great straits.'

In fact, John had borrowed money in order to buy William the plantation. The plantation was a failure and a chastened Billy, reportedly sick and nearly starving, had to be rescued by Henry Laurens, a friend of his father, a year later. Laurens wrote to John that Billy 'lacked resolution', though he thought him a 'worthy, ingenious man'. Laurens went on to say: 'No colouring can do justice to the forlorn state of poor Billy Bartram. A gentle, mild young man, no wife, no friend, no companion, no human inhabitant within nine miles ... scant of the bare necessities, void of all the comforts in life, save an inimitable degree of patience.' A description to move a father's heart (it is possible Billy even encouraged Laurens to write about him with such pathos) and John Bartram soon bailed out Billy, once again.

'The greatest natural botanist in the world'

John Bartram can with accuracy be called the father of botany in North America. Bartram was born on a farm outside Philadelphia in 1699. Growing up a Quaker, he loved anything to do with nature, and wanted to become a doctor so he could closely study herbs. His father was a farmer, however, and so John decided to pursue that way of life. At the age of twenty-nine, he bought a farm and married, then remarried after his first wife died, and sired eleven children. His son William was born when John was forty.

Through the hustle and bustle of his life, John studied natural things. In the fall, when the crops had been harvested, he set out on jaunts to find plants, flowers and seeds. Unlike the frailer William, John was a tall and robust man. He wandered the wilderness of America, travelling to Virginia, North Carolina and Florida, often alone and in danger from Indians, a tough life for a man whose own father had been killed by Cherokees in 1709. John Bartram did most of his collecting for a London merchant, Peter Collinson, who so appreciated the unique plants Bartram found that he induced as many as fifty members of British royalty and nobility—including the Prince of Wales—to pay Bartram to collect for them. At the time, manicured 'English' gardens had gone out of style and most wealthy people wanted 'natural' gardens, with a Rousseau-like profusion of trees, overhanging bushes and irregular paths. There is a possibly apocryphal story that some of Bartram's cuttings contributed to the beginning of the Revolutionary War, since the Prince of Wales refused to come in from the rain as he was watching them being planted and died of pneumonia, thus leaving the throne to his brother, who would become King George III.

In between such collecting jaunts—and being named Royal Botanist to the American colonies—Bartram cultivated, outside Philadelphia, a botanical garden that was the best in America and one of the best in the world, filled with hectares of wild trees and flowers (this is the same garden that would be visited by David Douglas in the early nineteenth century, see page 168). The garden would figure in Bartram's death, for he died in 1778, worrying that the British troops advancing on Philadelphia would harm his precious plants. They did not, and even the work of his son William could not overshadow the reputation of the man Carl Linnaeus called 'the greatest natural botanist in the world'.

Into the wild

By 1772 William Bartram was thirty-three years old and had almost nothing whatsoever to show for it. After his plantation failed, he returned to Philadelphia, where he went deeply into debt in a business enterprise and so decided to flee back to his uncle's trading post in North Carolina. His only stroke of good fortune was that Peter Collinson, still impressed by his sketches and watercolours, had gained him some commissions from wealthy British clients—the duchess of Portland, for instance, was paying Bartram to draw 'all Land, River and Sea Shells, from the very least to the greatest'. Another client, the Quaker naturalist Dr John Fothergill, had hired him to do a series of sketches of molluscs and turtles.

While William Bartram's writings afford few glimpses into his personal life—he never married, wrote few letters and threw away the few he received—he appears to have undergone an epiphany around this time, for, with heretofore undisplayed determination, he approached Dr Fothergill and asked him if he would finance him on a journey—this time a solo one—through the southern wilds to Florida. Fothergill kindly wrote to John Bartram—this shows once again how much in the shadow of his father William was—that 'for [William's] sake as well as thine, I shall be glad to assist him. He draws neatly; has a strong relish for Natural History; and it is a pity that such genius should sink under distress.'

So Billy Bartram had one last chance. After returning to Philadelphia to prepare for his voyage, he travelled by sea to Charleston, South Carolina, in the spring of 1733 and went from there via a small schooner to Savannah, Georgia. Safely arrived at Savannah, he spent forty pounds of the money Dr Fothergill had given him on a good horse—quite a sum at the time but a necessary outlay in Bartram's view, given the rigours of the travels ahead of him. Little did he know that he would run through several more horses, for he was now embarking on a journey that would leave him in the wilderness for five years, exploring Georgia and eastern Florida in 1773 and 1774, making further explorations in Georgia and South Carolina in 1775, and spending most of 1776 and 1777 in western Florida.

Bartram was setting off into the swamps and nearly impenetrable forests to the west of the thin strip of inhabited land on the coasts of these British colonies which, as the War of American Revolution broke out in 1776, would soon become battlegrounds. As he rode inland from Savannah, he described the changing scenery:

> From the sea coast, fifty miles [80 kilometres] back, is a level plain, generally of a loose sandy soil, producing spacious high forests ... Nearly one third of this plain is what the inhabitants call swamps, which are sources of numerous small rivers and their branches ... twenty or thirty miles [30–50 kilometres] upwards from the sea, when they branch and spread abroad like an open hand, interlocking with each other, and forming a chain of swamps across the Carolinas and Georgia, several hundred miles parallel to the sea coast ... [Finally the land] rises to a bank of considerable height ... which is mostly a forest of the great long-leaved pine, the earth covered with grass ...

During this five-year period, Bartram would appear periodically on the coast, to send specimens and drawings to sponsors, write his father a letter and then plunge back into the wilderness. Most of America and the world would not hear of his adventures until 1791, when he published his justifiably acclaimed *Travels through North & South Carolina, Georgia, East & West Florida, the Extensive Territories of the Muscogulges or Creek Confederacy, and the Country of the Chactaws* ...

'A restless spirit of curiosity'

Although he had been described as possessing a 'tender and delicate frame' and although he was prone to illness, Bartram was tough enough to live, literally for years, in the wilderness in a tent, occasionally pitched in the pastures of a remote farmer but mostly by himself, surrounded by the howl of cougars and the bark of foxes. There are times when he compared himself to the biblical figure Nebuchadnezzar, who had been thrown out to wander in the wilderness, but one also realises that Bartram relished his solitude and, in fact, thrived in it. 'Continually impelled by a restless spirit of curiosity, in pursuit of new productions of nature,' as he wrote, Bartram wandered on horseback through swamps whose trees dripped Spanish moss, through endless savannahs, along nameless rivers and through land dotted with strange mounds that he characterised (correctly) as having been left behind by an ancient civilisation of Indians who had vanished.

Like Alexander von Humboldt after him (see page 95), Bartram came to believe that creation was not merely a hierarchical pyramid with humans at the top, but an interwoven fabric of plants, animals and human beings, with all living things dependent on each other. His writings are from the heart of a person who feels more at home in the wilderness than he does with other people. One evening in Florida, he refreshed himself by eating a grilled trout he had caught in a stream, garnished with an orange he had plucked from a tree. He decided to save the rest of the fish for the next morning, and so hung it on an overhead branch of a shrub and made his bed near a fire.

All now silent and peaceable, I suddenly fell asleep. At midnight I awake; when, raising my head erect, I find myself alone in the wilderness of Florida, on the shores of Lake George ... I rekindle my sleepy fire; lay in contact the exfoliated smoking brands damp with the dew of heaven.

When quite awake, I started at the heavy tread of some animal; the dry limbs of trees upon the ground crack under his feet; the close shrubbery thickets part and bend under him as he rushes off.

The bright flame ascends and illuminates the groves around me.

[Waking in the morning] I found my fish carried off, though I had thought them safe on the shrubs, just over my head ... Perhaps it may not be time lost, to rest a while here, and reflect on the unexpected

and unaccountable incident, which however pointed out to me an extraordinary deliverance or protection of my life, from the rapacious wolf that stole my fish from over my head.

How much easier and more eligible might it have been for him to have leaped upon my breast in the dead of sleep, and torn my throat, which would have instantly deprived me of life, and then glutted his stomach for the present with my warm blood ...

Bartram marvelled at the fact that the wolf had decided to spare his life and to take only the more modest meal of the fish. The naturalist believed that animals had emotions and the ability to weigh such decisions. One day as he hiked, he was startled by a rattlesnake 'about six-foot [1.8 metres] in length and as thick as an ordinary man's leg'. He killed it but was instantly regretful: 'He certainly had it in his power to kill me almost instant, and make no doubt he was conscious of it ... I promised myself that I would never again be accessory to the death of a rattle snake, which promise I have invariably kept to.'

'A scene new and surprising'

Despite the closeness that Bartram felt when it came to animals, one of his most famous passages in *Travels* concerns a horrifying attack by alligators as he was travelling down the St John's River in Florida, scene of previous adventures with his father. The day was blazing hot, and Bartram was tired and discouraged because his Indian guide had run off, leaving him alone. Landing on a point of land for the night, he gathered dry wood—a rather difficult task—in order to build a smoky fire that might ward off the omnipresent mosquitoes. With that fire smouldering, he cast a line in the water, hoping to snag a trout.

But suddenly one alligator appeared, and then another, the two of them rivals locked in battle, which Bartram describes:

Behold [the alligator] rushing forth from the flags and reeds. His enormous body swells. His plaited tail brandished high, floats upon the lake. The waters like a cataract descend from his opening jaws. Clouds of smoke issue from his dilated nostrils. The earth trembles with his thunder. When immediately from the opposite coast of the lagoon emerges from the deep his rival champion. They suddenly dart upon each other. The boiling surface of the lake marks their rapid course, and a terrific conflict commences. They now sink to the bottom folded together in horrid wreaths.

This scene of internecine conflict was only a sample of what awaited Bartram, because soon he realised that alligators were advancing on him from all sides of the small peninsula he had camped on. He loaded his rifle but decided that his main line of defence should be a club, since the gun was capable of only one shot at a time. Leaping into his canoe, he attempted to paddle away.

I kept strictly on the watch, and paddled with all my might towards the entrance of the lagoon, hoping to be sheltered there from the multitude of my assailants; but ere I had half-way reached the place, I was attacked on all sides, several endeavoring to overset the canoe. My situation now became precarious to the last degree: very large ones attacked me closely, at the same instant, rushing up with their heads and part of their bodies above the water, roaring terrible and belching floods of water over me. They struck their jaws together so close to my ears, as almost to stun me, and I expected every moment to be dragged out of the boat and instantly devoured.

Certain that the canoe would capsize, Bartram decided to return to land, but the alligators followed him back. One of them got so close that Bartram stuck his gun against the animal's head and blew its brains out. After fighting off another alligator that appeared to be hunting in his canoe for fish, Bartram reconnoitred the land but found himself almost entirely surrounded by swamps and deep water. He had just about decided that his only hope of salvation lay in climbing a tree, when he heard a tumultuous noise from the river. Returning, his eyes beheld 'a scene new and surprising, which at first threw my senses into such a tumult that it was some time before I could comprehend what was the matter'.

Bartram had stumbled upon an alligator feeding frenzy. At certain times of the year on the St John's River, the trout descend in huge shoals and these alligators had come to greet them. The sight was astounding and terrifying—literally hundreds of alligators 'so close together from shore to shore, that it would have been easy to have walked across on their heads, should the animals have been harmless' stirring up the river into a red froth that looked like bloody rapids as they fed their insatiable appetites with thousands of fish.

For the next several days, Bartram was obliged to canoe down the river between banks filled with these huge creatures, which would sometimes swim out to attack him. Beating them off with clubs, he kept on going. One of his most famous sketches depicts the great creatures during this encounter, but after *Travels* was published the account was met with a great deal of scepticism, which wounded the naturalist greatly. In fact, recent behavioural studies have borne out Bartram's observations of these alligators.

'Beautiful and entertaining beings'

On a more peaceful note, Bartram loved to draw and write about birds. He loved the beauty of birds and their lovely song—'the sweet enchanting melody'—but also recorded their movements with a scientist's eye. At the time, the migration of birds with the seasons was not a well-observed phenomenon. In fact, wrote Bartram, 'even at this day very celebrated men have asserted that swallows, at the approach of winter, voluntarily plunge into lakes and rivers, descend to the bottom, and there creep into the mud and slime' where they hibernated for the winter.

Bartram admitted that he did not know how far south birds migrated, but he was quite certain that they did and then returned north in the spring. In any event, they were 'beautiful and entertaining beings', and that alone was reason enough to study them. Bartram placed, at the end of *Travels*, a listing of some 215 avian species. He was one of the first American naturalists to compile such a list, although, it turned out, he had identified many of them incorrectly, or at least not in strict accordance with the European methods of classification in use at the time—these changed often and in any event Bartram was not familiar with them. This was a problem for Bartram when it came to identifying other species as well. Many reptiles, fish and plants that he wrote of in *Travels* were later claimed as discoveries by other naturalists who knew the correct nomenclature.

Aside from plants and animals, Bartram was fascinated by Indians. Raised as a Quaker, he was more willing than many early Americans to accept Indians as human beings on a par with Europeans, and during his five years of travelling he developed a true admiration for the Indians he encountered, especially the Creeks, who were being gradually evicted from their ancestral homelands. Without, for the most part, succumbing to the myth of the noble savage, he admired their woodcraft, their customs and their freedom. The Creeks, for their part, got used to his strange, slow progress through their lands, bent over closely to the ground, examining things. They called him 'Plant Hunter'.

Bartram was also far more willing than most Europeans at the time to give the southeastern American Indians credit for having once been part of a larger and more powerful civilisation—one that, scientists have discovered, was destroyed by European diseases in the fifteenth century. One of Bartram's most haunting descriptions is that of finding, in the middle of the wilderness, a huge Indian mound—probably an earthen knoll used as a platform for the home of a priest-chief—surrounded by ancient and desolate fields. Extending from the mound 'was a noble Indian highway which led from the great mount on a straight line, three-quarters of a mile [1200 metres] ... and continued thence through an awful forest of live oaks.

This grand highway was about fifty yards [45 metres] wide, sunk a little below the common level, and the earth thrown up on each side, making a bank about two feet [60 centimetres] high.'

But even in Bartram's time, farmers were beginning to encroach on this scene from the distant past, and it was soon to disappear.

Homeward bound

Bartram's travels had left him weak with illness—at one point he became partially blind, possibly from a bout of scarlet fever—and so he decided to return home to Philadelphia, possibly impelled by the fact that the American Revolution was raging close to the home of his family. (As a Quaker pacifist, Bartram had taken little part in the war, although he had temporarily enlisted in Florida in a Patriot army that had been hastily gathered 'to repel a supposed invasion ... from St Augustine by the British', an invasion that never occurred.)

Bartram arrived back in Philadelphia in January of 1777, to find his father, aged seventy-eight, in poor health and fretting over the damage a British takeover of the city might do to his beloved botanical garden. John Bartram died in September of that year; it was no doubt a blessing for him to be able to see his 'indolent' son come home from the wilderness a man at last. After his father died, William Bartram lived in the family home and, going into partnership with his brother Moses—who handled the business side of things— managed the botanical garden. He did not publish his travels until 1791, fourteen years after his return, but when the book finally came out it went through numerous editions.

By the early nineteenth century Bartram was much sought after by scientists and philosophers (including Thomas Jefferson) and had been offered the first professorship of botany at the University of Pennsylvania (he never actually taught there, possibly because of his lack of formal education). His influence spread even to the arts. Bartram scholar Judith Magee writes: 'His lyrical descriptions of nature [in *Travels*] found great favor with European Romantic poets such as Samuel Taylor Coleridge and William Wordsworth, who borrowed heavily from his work in their poetry.'

William Bartram—Billy no more—died at the age of eighty-five in 1823, collapsing from a heart attack while partaking in a 'morning survey' of his beloved botanical gardens.

The Franklin tree

One day in 1765, while they journeyed along the river Altamaha in Georgia— 'how gently flow thy peaceful floods, O Altamaha', William would rhapsodise in Travels—*John and William Bartram came upon a grove of lovely small trees that they realised constituted a new species. Of them, William wrote:*

> It is a flowering tree of the first order for beauty and fragrance of blossoms; the tree grows fifteen or twenty feet [4.5–6.0 metres] in height ... the flowers are very large, expand themselves perfectly, are of a snow white color, and ornamented with a crown or tassel of a gold-colored refulgent staminae in their center.

They called the tree Gordonia Pubescens Franklinia, *or the Franklin tree, after John Bartram's good friend, inventor and statesman Benjamin Franklin (it is now known as* Franklinia altamaha*). When William returned to the region in 1773, he found the same grove of trees and took seeds back with him to the Bartram botanical gardens outside Philadelphia. In doing so, he probably saved the Franklin tree from complete extinction, because by 1803 this grove had disappeared from the wild and no other was ever found. Scientists disagree as to what happened. It may be that the grove of Franklin trees was established during the last Ice Age and was unable to withstand the more temperate climate, or possibly it was not genetically diverse enough to survive. And the Franklin tree is notoriously finicky and hard to grow, which would not improve its chances in the wild.*

In any event, the only reason it survives today as a popular ornamental tree is that it was saved by the Bartrams.

JOSEPH BANKS
Gentleman naturalist

Joseph Banks Esq, Fellow of the Society, a Gentleman of large fortune, who is well versed in natural history, being Desirous of undertaking the same voyage the Council very earnestly requests their Lordships, that in regard to Mr Banks's great personal merit, and for the Advancement of useful knowledge, He also, together with his Suite, being seven persons more, be received on board the ship, under command of Captain Cook.

This epistle, dated 15 February 1768, was a typically indirect yet elegant eighteenth century way for the writer of the letter—the scientific Royal Society of London—to inform the British Admiralty that they wanted a man named Joseph Banks to accompany Captain James Cook on what was sure to be a historic voyage to the South Seas. The salient information was that Banks was a gentleman 'of large fortune', which meant that he could afford to fund his own passage on the expedition, along with his 'suite' of seven servants and fellow scientists, not to mention his dogs.

The Admiralty acceded to this request—Banks would fund a great deal more of the voyage than his own passage and that of his friends— perhaps thinking that Banks was another of the gentleman dabblers in science who, popular imagination had it, populated the Royal Society (although the society also included the likes of Sir Joshua Reynolds, Voltaire, Carl Linnaeus, and numerous other genuine luminaries). But Banks was far more than a gentleman dabbler. As his three-year voyage to the South Seas and beyond with Captain James Cook would prove, he was one of the finest naturalists of his age, a man who would put his vast fortune to the service of learning.

'How beautiful!'

Joseph Banks was born on 13 February 1743, with a silver spoon gleaming in his mouth. His father, William Banks, was a lawyer and member of the House of Commons who had inherited a vast family fortune and upon whose Lincolnshire estate Banks, along with his sister, grew up. Banks was not a studious young man; he attended Harrow, Eton and then Oxford but, like so many naturalists, really preferred wandering the fields, studying plant life and animals. As reported in Patrick O'Brian's biography of Banks, one friend wrote that Banks 'cared mighty little for the book'. Another said that Banks had had a transformation at Oxford while coming back from swimming on a summer evening:

> *He was walking leisurely along the lane, the sides of which were richly enameled with flowers; he stopped and looked round, involuntarily exclaimed, How beautiful! After some reflection, he said to himself, it is surely more natural that I should be taught to know all these productions of nature, in preference to Greek and Latin ...*

In 1761, when Banks was only eighteen, his father died and Banks began to divide his time between London, his family estates and Oxford, although he left the university in 1765 without receiving a degree. He had come into a vast fortune when he turned twenty-one the previous year and might have become another gentleman living off the interest of the family estates while stuffing dried flowers between the leaves of books, but he was too ambitious and curious for that. While managing the family money, he devoted himself to pursuing the natural sciences. Establishing his own home in London, he began to entertain many of the brilliant botanists of his day, including Philip Miller, the distinguished head gardener of the Physic Garden in Chelsea and a friend of Carl Linnaeus, the Swedish botanist whose naming system was to revolutionise botany (see page 36). Another connection to Linnaeus was Linnaeus's student Daniel Solander, a Swede who had travelled to Great Britain to promote Linnaeus's system of classification and met Banks while at the British Museum.

Solander, ten years older than Banks, would become fast friends with the English naturalist, even though Banks habitually referred to him with the formal title of 'Dr Solander'. Both men were modest and extremely 'amiable', in the parlance of the day, and both were anxious to strike out on their own into the field. Banks would go first, with a voyage to Newfoundland and Labrador.

'Seven Islands of Ice'

Banks was twenty-three years old when he set sail for Canada aboard HMS *Niger*, in April of 1766. The *Niger* was headed to Newfoundland to patrol the rich British cod-fishing grounds off the coast. Accompanied by an old school friend, Lieutenant Constantine John Phipps—a sailor, but also a mathematician and astronomer—Banks was embarking on his first naturalist's adventure. His journal, with its idiosyncratic punctuation and spelling, is filled with the observations of a burgeoning scientist:

> *May 3 Today [off the coast of Newfoundland] a calm fish'd with a landing net out of the Quarter Gallery window. Caught see weed, Fucus acinarius, with fruit like Currants on slight Footstalks ... also two species of what the seamen call Blubbers [jellyfish] one rounded and Transparent with edges a little Fringed ...*

May 9 This Morn Seven Islands of Ice in sight one Very Large but not high about a League from us we steer very near a small one which from its Transparency & the Greenish Cast in it makes a very Beautiful appearance ...

Banks soon supplemented his notes on plant life with his observations on the Indians he met. In August, he travelled with Phipps to Labrador, where he found a wrecked birch-bark canoe, which was, he thought, a sign that the local Newfoundland Indians were nearby. He did not meet them but had heard about them from British fishermen, with whom they were in 'a continual State of Warfare'. Whether the notations on the luckless Sam Frye below are completely accurate, they show a young naturalist with a careful eye for detail and flair for colour:

[The Indians] Canoes by the Gentlemans account from whom I have all this are made like the Canadians of Birch Bark sewd together with Deers sinews or some other material but Differ from the Canadians Essentially in that they are made to shut up by the sides Closing together fro the Convenient Carriage [on portages through the woods] ...

Their method of scalping to is very different from the Canadian they not being content with the hair but skinning the whole face at Least as far as the upper Lip.

I have a scalp of this Kind which was taken from one Sam Frye a fisherman who they shot in the water last year but the features were so well Preservd that when upon a party of them being Pursued the next Summer they Dropd it it was immediately Known to be the scalp of the Identical Sam Frye ...

Banks stayed in Canada until October, when the *Niger* sailed for Lisbon, along the way hitting 'a very hard Gale of Wind' that destroyed some of the specimens he had painstakingly gathered. Nonetheless, by the time he made his way back home in January 1767,

Banks had with him 340 plant specimens and over ninety preserved birds (collected on behalf of the Welsh naturalist Thomas Pennant). Banks had numerous artists, including Sydney Parkinson, illustrate selected items from his collection. While not as glorious as the one Banks would ultimately host, this collection of specimens impressed many at the Royal Society and was instrumental in getting him an invitation on one of the greatest sea voyages of all time.

The *Endeavour* sets sail

As it turned out, another soon-to-be-famous Englishman was working in the waters off Newfoundland while Banks was there—a junior officer in the Royal Navy named James Cook, who was charting the previously unmapped Canadian coastline as well as observing the solar eclipse of the sun that occurred in August 1766. His work in Canada brought him to the attention of the Royal Society, which was planning to send an expedition to the island of Tahiti (recently discovered by Captain Samuel Wallis of the Royal Navy) in order to observe the Transit of Venus, in which Venus passes across the face of the sun—something that happens only once in 243 years. Observing this phenomenon from different parts of the globe, it was thought, would help measure the distance from the earth to the sun and facilitate navigation for a nation whose seagoing fortunes were on the rise.

Cook, promoted now to lieutenant, was tapped by the Royal Navy to command HMS *Endeavour*, a former coal shuttle transformed into a home for eighty-five sailors, marines and officers—and Joseph Banks and his guests. Banks had, in fact, bankrolled a good part of the voyage, putting down even more money than King George III, whose pet project the venture was. Government interest was high because, as well as observing the Transit, Cook had secret orders— after leaving Tahiti, he was to find the continent known as *Terra Australis Incognita*, the Unknown Southland, the southern continent that had long been rumoured to exist in the Pacific Ocean somewhere between New Zealand and South America.

In August of 1768, Banks, four servants, three artists (including Sydney Parkinson) and the botanist Daniel Solander set sail aboard the *Endeavour*. They headed down the Atlantic to the coast of Africa and then sailed diagonally across the ocean, making the famous Atlantic Passage, before arriving in Brazil in November. After provisioning, the *Endeavour* sailed down the South American coast. As they proceeded, Banks could feel the temperature cooling 'and shut two of the Cabbin windows, which have been open since Madeira'. They were heading into the swells of the South Atlantic, which were frigid but nonetheless full of life—different types of plankton (Banks saw them as 'broad swathes of particles' in the water), many whales and numerous types of birds. On 22 December Banks wrote:

> *This morn quite calm. A very large shoal of Porpoises came close to the ship, they were of a kind different from any I have seen but so large that I dared not throw the gig into any of them, some were 4 yards [3.6 metres] long, their heads quite round but their hinder parts compressd, they had one fin upon their backs like a porpoise and white lines over their eyes also a spot of white behind the fin; they stayd above an hour about the ship. When they were gone Dr Solander and myself went out in the boat and shot one species of Mother Careys chickens and two shearwaters, both provd new,* Procellaria Gigantea *and* sandaliata. *The Carey was one but ill describd by Linnaeus,* Procellaria fregata. *While we were out the people were employed in bending the new set of sails for Cape Horn.*

The ship accomplished the rounding of Cape Horn by the end of January, after Banks and a few others stopped to visit the miserable inhabitants of Tierra del Fuego, and then headed into the vast reaches of the Southern Ocean, heading northwest. The peaceful passage was marred by an incident in which one of the young marines on board was accused of stealing a piece of sealskin from one of Captain Cook's servants. It was a small matter and the servant, recovering the sealskin, said he would not report it, but the young marine's fellow soldiers reproved him for it, saying he had blackened the honour of the squad, and Banks, with typical empathy, recalls what happened:

> *This even one of our marines threw himself overboard and was not miss'd till it was much too late even to attempt to recover him. He was a very young man scarce 21 years of age, remarkably quiet and industrious, and to make his exit the more melancholy was drove to the rash resolution by an accident so trifling that it must appear incredible to every body who is not well accquainted with the powerfull effects that shame can work upon young minds.*

However, even this horror was soon forgotten when the *Endeavour* arrived at the mountainous isle of Tahiti on 13 April.

'With a fire in her eyes'

The first European to arrive on Tahiti—o'Tahiti, as its inhabitants called it—had been Samuel Wallis, two years before the *Endeavour*, and Cook was to find out that the Frenchman Louis Antoine de Bougainville had also landed there the year before. Wallis had called the place King George's Island, a name that refused to stick to this paradise. Tahiti was a wonderfully rugged and mountainous island, covered with thick vegetation, with waterfalls roaring down the sides, a verdant land that had been inhabited by Polynesian voyagers ever since AD 800, if not before. Its inhabitants were handsome, intelligent, sexually liberated and warlike (at least among themselves). They were also voracious thieves and soon the chief occupation of most of the British officers and sailors visiting the island was to keep from being robbed blind, especially of iron, which was among the highest treasures the average Tahitian could imagine. British sailors began to trade ship's nails for sex; Cook frowned upon this and soon drew up a list of 'trading rules'—infractions were punishable by flogging.

Banks was busy helping search out a proper location to build an observatory for Cook's scientists to observe the Transit of Venus, which would occur on 3 June, but he was not immune to the charms of Tahitian ladies, as this entry from his journal shows:

> *Our cheifs own wife (ugly enough in conscience) did me the honour with very little invitation to squat down on the mats close by me: no sooner had she done so than I espied among the common croud a very pretty girl with a fire in her eyes that I had not before seen in the countrey. Unconscious of the dignity of my companion I beckond to the other who after some intreatys came and sat on the other side of me: I was then desirous of getting rid of my former companion so I ceas'd to attend to her and loaded my pretty girl with beads and every present I could think pleasing to her: the other shewd much disgust but did not quit her place and continued to supply me with fish and cocoa nut milk.*

The incident was left unresolved when Dr Solander had his pocket picked of a snuffbox and the dinner broke up, but one can see Banks (only twenty-four, after all) enjoying himself a good deal. In truth, he cared little for the Transit of Venus, as he intimated to friends. Thus, when the Transit turned out to be a disappointment—turbulence in the earth's atmosphere threw off the instruments of the scientists— and the *Endeavour* shortly thereafter left the island, the real journey, as far as Banks was concerned, was only beginning.

'We came up with it very slowly'

The *Endeavour* left Tahiti soon after observing the Transit, intent on Cook's secret mission—to find *Terra Australis Incognita*, if such a continent existed. Cook sailed among the Society Islands, helped in good part by a Tahitian priest named Tupaia, who had come along with the British. Cook searched for the landmass of the southern continent, heading as far down as the stormy latitudes of what would become known to a later generation of sailors as the Roaring Forties.

Here the waves were so mountainous that it was obvious there was no continent-sized land to impede them, and so Cook finally pointed the *Endeavour* west, seeking the land (present-day New Zealand) that the Dutch navigator Abel Tasman had found a century earlier.

Banks was in his element—one imagines him standing by the ship's rail, Solander by his side, pointing out birds in flight and fish broaching the ocean waves. On 19 September, he wrote:

> *Quite calm today go out in the boat and shoot* Procellaria velox *(the dove of the 31st),* Vagabunda *(the grey backd shearwater of the same day),* Passerina *(the small Mother Careys chicken of the 10th). Took with the dipping net* Medusa vitrea, Phillodoce velella *to one specimen of which stuck* Lepas anatifera, Doris complanata, Helix violacea, Cancer ... *Very few birds were to be seen, there were however some Albatrosses and a kind of Shearwater quite black which I was not fortunate enough to shoot. A large hollow swell from the South.*

And, on 6 October 1769, Banks and the rest of the crewmembers turned their heads in excitement as 'a small boy who was at the mast head Calld out Land'. It was New Zealand. 'We came up with it very slowly,' Banks wrote. 'At sun set myself was at the masthead, land appeard much like an Island or Islands but seemd to be large. Just before a small shark was seen who had a very piked nose something like our dog fish in England.'

They had arrived on the east coast of New Zealand's North Island, the first white men to approach the land since Tasman. Cook turned the *Endeavour* south and within a few days made landfall. Banks describes their first encounter with the Maori who inhabited the country:

> *In the evening went ashore with the marines &c. March from the boats in hopes of finding water &c. Saw a few of the natives who ran away immediately on seeing us; while we were absent 4 of them attacked our small boat in which were only 4 boys, they got off from the shore*

in a river, the people follow'd them and threatned with long lances; the pinnace soon came to their assistance, fir'd upon them and kill'd the cheif. The other three dragg'd the body about 100 yards [90 metres] and left it. At the report of the musquets we drew together and went to the place where the body was left; he was shot through the heart. He was a middle siz'd man tattow'd in the face on one cheek only in spiral lines very regularly form'd; he was cover'd with a fine cloth of a manufacture totaly new to us, it was tied on exactly as represented in Mr Dalrymples book p.63; his hair was also tied in a knot on the top of his head but no feather stuck in it; his complexion brown but not very dark.

It was no surprise blood was shed almost immediately—the Maori, descended from Polynesians who had sailed to the island around AD 800, had every reason to assume the British were a war party of territorial males. Every encounter the crew of the *Endeavour* would have with the Maori was at least initially hostile, until British weapons proved their superiority and the Maori were temporarily cowed. In the face of this, it is remarkable that Banks continued to gather specimens, but he did. Leaving behind Poverty Bay—so Cook had named the site of their first encounters with the Maori—the *Endeavour* sailed down the east coast of North Island until it reached turbulent waters at the southern end of the island. There, at what became known as Cape Turnagain, Cook decided to head back up the coast, and thus began his 4000-kilometre (2500-mile) circumnavigation of New Zealand.

By January, Cook had turned the North Cape of New Zealand and was coasting down the western side of the island, with Banks noting the hilly nature of the land:

The countrey ... appeard very pleasant and fertile, the sides of the hills sloping gradualy; with our glasses we could distinguish many white lumps in companies of 50 or 60 together which probably were either stones or tufts of grass but bore much the resemblance of flocks of sheep.

At night a small fire which burnd about an hour made us sure that there were inhabitants of whoom we had seen no signs since the 10th.

On 17 January the *Endeavour* turned to shore again, and Banks wrote in his journal that he and the rest of the shore party found a family making their dinner near a basket filled with human bones:

... the bones were clearly human, upon them were evident marks of their having been dressd on the fire, the meat was not intirely pickd off from them and on the grisly ends which were gnawd were evident marks of teeth, and these were accidentaly found in a provision basket. On asking the people what bones are these? they answerd, The bones of a man.—And have you eat the flesh?—Yes.—Have you none of it left?—No.—Why did not you eat the woman who we saw today in the water?—She was our relation.—Who then is it that you do eat?—Those who are killd in war.—And who was the man whose bones these are?— 5 days ago a boat of our enemies came into this bay and of them we killd 7, of whoom the owner of these bones was one.—The horrour that apeard in the countenances of the seamen on hearing this discourse which was immediately translated for the good of the company is better conceivd than describd.

Banks was more amused than horrified and even purchased— with a pair of his linen drawers—the head of one of the victims, writing (ever the scientific observer) that 'the flesh and skin were soft but they were somehow preserved so as not to stink at all'.

When Cook had finished sailing down the west coast of North Island, he headed east through what is now known as Cook Strait, then south around South Island, navigating it clockwise. In the course of six months, he had completed a figure-eight circumnavigation of New Zealand, and he decided to head home, sailing west, intending to round Africa's Cape of Good Hope. He had not counted on one thing, however: Australia, whose coastline he raised on 19 April 1770.

'His house is a perfect museum'

Joseph Banks returned from his Endeavour *voyage with innumerable natural specimens, which, in the latter half of 1772, he had arranged in his London home on New Burlington Street. A gentleman who visited him there in December of that year was suitably impressed, as he wrote to a friend:*

[Banks's] house is a perfect museum; every room contains an inestimable treasure. I passed almost a whole day there in the utmost astonishment, could scarce credit my senses ... [In one room] there is a large collection of insects, several fine specimens of the bread and other fruits preserved in spirits; together with a compleat *hortus siccus* [Latin for 'dried garden', a collection of dried plants] of all the plants collected in the course of the voyage. The number of plants is about 3,000, 110 of which are new genera, and 1300 new species which were never seen or heard before in Europe ... [Another room] contains an almost numberless collection of animals: quadrupeds, birds, fish, amphibia, reptiles, insects and vermes, preserved in spirit, most of them new ... Here I was most in amazement and cannot attempt any particular description. Add to this the choicest collection of drawings in Natural History that perhaps ever enriched any cabinet, public or private—987 plants drawn and coloured by [Sydney] Parkinson ... and what is more extraordinary still, all the new genera and species contained in this vast collection are accurately described, the description fairly transcribed and fit to be put to press.

Strangely enough, despite Parkinson's drawings and the work of Dr Solander—for it was he who did the laborious work of writing the scientific descriptions for each of Parkinson's drawings—Banks never published the results of his voyage, for reasons scholars have attributed to everything from a lack of focus to an insecurity about his own work. His herbarium collection now resides in the Botany Department of the Natural History Museum, in London.

'A most terrible surf'

Banks never expresses a great deal of emotion or surprise in his journal. Finding land where none was supposed to be, he merely wrote: 'With the first day light this morn the Land was seen, at 10 it was pretty plainly to be observd; it made in sloping hills, coverd in Part with trees or bushes, but interspersd with large tracts of sand. At Noon the land much the same.' He was more interested in the almost hallucinatory waterspouts that played across the ocean, 'about the thickness of a mast or a midling tree' and 'reachd down from a smoak colourd cloud about two thirds of the way to the surface of the sea'.

Banks was certainly not impressed by the land the *Endeavour* sailed north along: 'The countrey ... resembled in my imagination the back of a lean Cow, coverd in general with long hair, but nevertheless where her scraggy hip bones have stuck out farther than they ought accidental rubbs and knocks have intirely bard them of their share of covering.' And he was most certainly not impressed by the people: the Aboriginals appeared strangely incurious about these new people in their huge boat and, aside from a few skirmishes, for the most part melted away.

But when Cook finally landed on the shores of what he initially called Sting Ray Harbour, because of the vast numbers of those creatures present, Banks and Solander made repeated expeditions ashore, finding themselves in a kind of naturalist's paradise. On 3 May Banks wrote:

> *Our collection of Plants was now grown so immensly large that it was necessary that some extrordinary care should be taken of them least they should spoil in the books. I therefore devoted this day to that business and carried all the drying paper, near 200 Quires of which the larger part was full, ashore and spreading them upon a sail in the sun kept them in this manner exposd the whole day.*

The next day he wrote: 'Myself in the woods botanizing as usual, now quite void of fear as our neighbours have turnd out such rank cowards.' He collected so many hundreds of plants that Cook renamed the harbour Botany Bay—as Cook wrote in his own journal: 'The great number of New Plants &ca Mr Banks and Dr Solander collected in this place occasioned my giveing it the name ...'

On 5 May they left Botany Bay and sailed north along the Australian coastline, charting and taking soundings, blissfully unaware that the Great Barrier Reef, protecting 2000 kilometres (1250 miles) of the coastline, lay ahead. Their idyll was soon over, however. Running aground on shoals, they were forced to land the *Endeavour* and spend seven weeks patching up its damaged hull. During this time, Banks, out gathering plants, saw a kangaroo: 'He was not only like a grey hound in size and running but had a long tail, as long as any grey hounds, what to liken him to I could not tell, nothing certainly that I have seen at all resembles him.' Soon Banks got close enough to shoot one of the creatures, killing one that weighed 38 kilograms (84 pounds).

After setting sail again in August, they discovered that they were 'embayed' by a giant reef. On 15 August Banks wrote:

> *A Reef such a one as I now speak of is a thing scarcely known in Europe or indeed any where but in these seas: it is a wall of Coral rock rising almost perpendicularly out of the unfathomable ocean, always overflown at high water commonly 7 or 8 feet, and generaly bare at low water; the large waves of the vast ocean meeting with so sudden a resistance make here a most terrible surf Breaking mountain high, especialy when as in our case the general trade wind blows directly upon it.*

Cook, showing fine seamanship, found a safe passage out of the reef but then was nearly blown back against the sharp coral by the prevailing trade winds. Luckily, he found what they named

Providential Channel, which led them back into sheltered waters. Cook finally sailed them around the northern tip of Australia and then westward, for home.

'At 3 O'clock Landed at Deal'

Joseph Banks returned home on 12 July 1771, landing at Deal, England, as his last journal entry reads. He brought back a great haul. In three years he and Solander had collected thirty thousand plants (1400 of them unknown to science) and a thousand animal specimens. Hundreds of these had been sketched and painted by the artist Sydney Parkinson, who had accompanied Banks on the voyage and who died at sea in January 1771 on the journey home.

Banks initially agreed to go along with Cook on his second voyage but ultimately decided that he would prefer to stay in England, working on his collection. He made a subsequent exploring expedition to Iceland, but for the most part elected to remain at home, where he had become famous after the *Endeavour* expedition. He became president of the Royal Society in 1778, a position he held for a remarkable forty-one years, married respectably in 1779 (although there were some intrigues with women) and became a close advisor to King George III, who made him a baronet in 1781. It was Banks who made the royal gardens at Kew famous throughout the world.

Banks—growing more corpulent with age, but always gentle, and extremely kind, especially to young naturalists—was responsible for a number of British expeditions in which botanists sought to find plants that would improve Britain's economic fortunes. William Bligh's ill-fated voyage to gather breadfruit in Tahiti and bring them to the West Indies was Banks's idea, and Banks also pinpointed Assam, India, as the perfect location to grow tea for export to Great Britain.

Suffering severely from gout, painful kidney stones and other ailments, Joseph Banks died in June of 1820, one of the most influential naturalists of his age.

'To prove that he was a Woman'

Joseph Banks was a wealthy, handsome, intelligent and kind person; the women he met certainly appreciated all these traits. His first serious love affair seems to have been with Harriet Blosset, a Frenchwoman living in England. The fact that Banks could speak no French did not seem to in any way retard the progress of this affair—one observer saw the two just before Banks boarded the Endeavour *and remarked that they seemed smitten with each other.*

However, when Banks returned from the voyage the ardour—on his part—had definitely cooled. According to a gossipy letter writer of the time, Banks, now famous, did not contact Harriet, and 'Miss Blosset set out for London and wrote him a letter desiring an interview of explanation. To this Mr Banks answer'd by a letter of two or three sheets professing love &c but that he had found he was of too volatile a temper to marry.' Rumour had it that Harriet Blosset's family was given a large sum of money so that they would not sue for alienation of affections.

In 1779 Banks married Dorothea Hugessen and remained married to her for the rest of his life, but between Harriet and Dorothea there was the curious matter of 'Mr Burnett'. In August of 1772, on his way again to the South Seas aboard the Resolution, *Captain James Cook stopped at Madeira Island. 'Eleven days before we arrived,' Cook wrote from Madeira, 'a person left the Island who went by the name of Burnett he had been waiting for Mr Banks arrival about three months, at first he said he came here for the recovery of his heath, but afterwards said his intention was to go [to the South Seas] with Mr Banks ... at last when he heard that Mr Banks did not go, he took the very first opportunity to get of the island ... Every part of Mr Burnets behaviour and eery action tended to prove that he was a Woman. I have not met a person who entertains a doubt of a contrary nature ...'*

Scholars and biographers have puzzled over this, but it seems fair to believe that 'Mr Burnett' was a lover of Banks who went to Madeira to meet Banks without the spying eyes they would have endured in England. (Banks, of course, decided against accompanying Cook a second time, and Mr Burnett, hearing this news, immediately decamped back to England.) Would Banks really have attempted to take her along on the journey? We will never know.

ALEXANDER
VON HUMBOLDT
'the unity of nature'

A telling moment in the life and career of Baron Alexander von Humboldt came in Paris, in 1804. Fresh from five years of extraordinary adventures in the Americas, Humboldt was the toast of Europe and one evening was introduced to another man who also held the Continent in thrall—Napoleon Bonaparte. 'So, monsieur,' Napoleon remarked. 'You've been studying Botanics?' Showing an unusual amount of modesty, Humboldt nodded in agreement. Napoleon sneered: 'Just like my wife.' And walked away.

It is not hard to figure out why Napoleon responded as he did—although Napoleon was one of the most famous men in Europe, Humboldt was *the* most famous. Everyone had heard of his travels and would soon read his *Personal Narrative of Travels to the Equinoctial Regions of the New Continent during the Years 1799–1804,* which told of his trek through Latin America, where he gathered forty-two crates filled with nearly six thousand specimens of plant, insect and animal species, many new to science—not a mere study of 'Botanics'. Humboldt also provided geological specimens—few had previously been brought back from Latin America—with hundreds of pages of field notes.

Just as the exploits of the short Corsican Napoleon would affect the future of world history, the travels of the tall, aristocratic, impossibly vain, probably homosexual, Prussian Humboldt would have an enormous impact on those who sought out wild places in order to study nature. Humboldt was the man after whom the young Darwin fashioned himself (Darwin called Humboldt 'the greatest travelling scientist who ever lived'), and it was Humboldt who created the model of the intrepid natural adventurer to be followed by the likes of David Douglas, Henry Bates, and many others. Plunging into jungles, climbing mountains, canoeing down strange rivers—and all the time noting, inspecting, sampling, collecting—Alexander von Humboldt was the naturalist's naturalist.

'I was anxious to contemplate nature'

'From my earliest youth,' Alexander von Humboldt wrote at the beginning of his *Personal Narrative*, 'I felt an ardent desire to travel into distant regions, seldom visited by Europeans.' Humboldt was born in 1769 into a wealthy Prussian family whose estates lay on the outskirts of Berlin. His father died when he was ten. His mother pushed for him to become a civil servant, but this held little appeal, and so he pursued a degree in mineralogy at the University of Göttingen and went on to train at the School of Mines in Freiburg before being awarded a position as a mining inspector by the Prussian government.

Humboldt discharged his duties well enough, and even conducted important experiments on identifying air quality in the mines, but his heart was not in it. In 1796, when Humboldt was twenty-seven, two things happened. His mother died of breast cancer, leaving him with an income of three thousand thalers per year, which meant that, in modern terms, he was quite comfortably a millionaire. And Reinhard von Häften, the young infantry lieutenant whom most scholars consider to have been the love of Humboldt's life ('I know that I live only through you,' Humboldt once wrote to him, 'and that I can only be happy in your presence'), decided to get married.

'A man should get used to standing alone early in life,' Humboldt wrote to a friend at the time. 'Isolation has a lot in its favour.' Throwing off his grief—or burying it—he decided to follow his early dream of exploring the 'distant regions' of the earth. 'I was anxious to contemplate nature in all her variety of wild and stupendous scenery,' he later wrote. 'The hope of collecting facts useful to the advancement of science, incessantly impelled my wishes toward the luxuriant regions of the torrid zone.'

At first, though, it seemed that Humboldt's wishes would be thwarted, despite his desire and deep pockets. He signed on for a 1798 British expedition to Egypt, only to have the trip cancelled because Napoleon's troops were just about to invade North Africa. He then accepted a position with the French explorer Louis Antoine de Bougainville, who was planning a five-year circumnavigation of the globe, but this trip was cancelled after Napoleon invaded Austria, forcing the French government to husband its resources for war. Impatiently, Humboldt said: 'A man can't just sit down and cry. He's got to do something.' Joining forces with another naturalist who had hoped to go on Bougainville's expedition, twenty-five-year-old Aimé Bonpland, Humboldt set off on his own expedition to North Africa. But, through a series of misadventures, he and Bonpland ended up in Spain, where Humboldt's aristocratic connections got them an audience with the king, who seized the opportunity to ask Humboldt if he—as a Prussian mining professional—might be able to inspect the king's mines in his colonies in the New World. On the spur of the moment, Humboldt agreed. Instead of heading for North Africa, he and Bonpland were now pointed to an adventure in the Americas.

'We beheld a verdant coast'
Humboldt's enthusiasm grew as he prepared for their journey. He understood that the Spanish colonies—which extended all through South America, Central America, Mexico, parts of what is now the southwestern United States and California—were mainly closed to

foreign travellers and had been for three hundred years. Although scientific expeditions had nibbled around the coastlines of the Spanish colonies, the interior of South America was a great unknown, from which issued wild legends. Yet here he had been given unprecedented access: 'Never had so extensive a permission been granted to any traveler, and never had any foreigner been honored with more confidence on the part of the Spanish government.'

Humboldt and his companion set sail aboard the frigate *Pizarro* in June of 1799, sailing under blackout conditions at night to avoid British fleets seeking to capture any ship leaving a Spanish port, since Spain was an ally of Napoleon. During the day, however, he and Bonpland measured winds, currents and water temperature, and Humboldt was amused that his sextant readings of the ship's positions were more accurate than those of the *Pizarro*'s captain. (The two young scientists carried with them more than forty instruments—rain gauges, thermometers, even a sextant that could fit inside a snuffbox.) When the *Pizarro* stopped to take on supplies at Tenerife, the largest of the Canary Islands, Humboldt and Bonpland took advantage of the break from shipboard life to climb Mount Peyde, a semi-dormant 3750-metre (12,300-foot) volcano. It was an experience of temperature extremes. Climbing to the summit of the mountain by scrambling up an old and frozen flow of lava, they found themselves, at eight in the morning, shivering from the cold, in a freezing wind, but when they dangled their feet over the volcano's smouldering caldera, they felt intense heat rising.

After a stop in Havana, the *Pizarro* arrived off the coast of Venezuela on 16 July. Even though typhus had struck the vessel, killing one crewmember and causing numerous others to sicken, Humboldt and Bonpland were impressed by their first glimpse of the mainland: 'At break of day we beheld a verdant coast, of picturesque aspect. The mountains of New Andalusia [Venezuela], half-veiled by mist, bounded the horizon to the south. The city of Cumaná and its castle appeared between groups of cocoa trees.'

The Humboldt Current

In December of 1802 Humboldt and Bonpland sailed up the western coast of South America as far as Mexico, carried along in part by the cool and swirling Pacific current known as the Humboldt, or Peruvian, Current.

The Humboldt Current's shallow, cold, low-salinity waters flow up from the Antarctic past the coasts of Chile and Peru, and then turn west. Deep, nutrient-rich water is left behind, creating an extraordinary eco-system—about twenty per cent of the world's fish come from these teeming and fertile waters, where small fish such as anchovies and sardines thrive and draw larger creatures such as dolphins, tuna and porpoises, as well as the sea birds that cover islands with huge deposits of guano, at one time so valuable for fertiliser that wars were fought over it.

In 1802 Humboldt visited one such guano island, Mazorca, which was covered with literally hundreds of metres of guano. (It was in part the samples of guano Humboldt brought home—Indians had used it for centuries as fertiliser—that brought it to the attention of European markets.) He found Mazorca incredibly humid, yet knew it had been without rainfall, literally for centuries. After measuring the ocean temperatures and finding them unusually cool, he realised that the Peruvian Current (as he called it) consisted of cool water, which combined with warm air to bring about the dense humidity that helped compact the guano. He also noted that when the cool air from the current blew over the mainland, it became warmer and was thus able to hold more moisture. The dense rainclouds thus created blow eastwards and pour out their contents at the base of the Andes.

'A fabulous and extravagant country'

The travellers had not intended to stop at Cumaná, but the typhus spreading through the crew and passengers of the *Pizarro* convinced them that now was the time to disembark from the vessel. It was a happy accident that they did. One of the very first cities founded by the Spanish on South American soil, Cumaná had an air of peace and orderliness, with cobblestone streets swept by breezes off the

Caribbean, although it was prone to occasional violent earthquakes. The governor, Don Vincente Emparan, welcomed Humboldt and Bonpland, professing himself to be 'much interested in anything that related to natural philosophy', and helped them rent a house that was large and had 'an agreeable coolness when the breeze arose'.

For the next several months, Humboldt and Bonpland toured the area with the gleeful satisfaction of children set loose in a sweet shop. 'What a fabulous and extravagant country we are in!' Humboldt wrote to his brother. 'Fantastic plants, electric eels, armadillos, monkeys, parrots: and many, many, real, half-savage Indians ... We've been running around like a couple of mad things; for the first three days we couldn't settle to anything: we'd find one thing, only to abandon it for the next. Bonpland keeps telling me he'll go out of his mind if the wonders don't cease.'

The two men were endlessly fascinated by the southern night skies, what Humboldt called in his *Personal Narrative* 'the brilliancy of the starry vault of heaven'. 'Venus plays the role of the moon here,' he went on to write. 'She shows big, luminous haloes and the most beautiful rainbow color, even when the air is quite clear and the sky is perfectly blue.' They measured the temperature of the water and air, counted the amount of salt produced in the salt mines and killed a guacharo bird, unknown to European science, which had an extra soft pad of fat around its chest that made for excellent cooking oil. They found, living in poverty among the mulattoes and Indians, an ancient Castilian man, a shoemaker—a trade, Humboldt remarks, that 'could not be very lucrative in a country where the greater part of the inhabitants go barefooted'. But the man turned out to be 'a sage of the plain; he understood the formation of the salt by the influence of the sun and full moon, the symptoms of earthquakes, the marks by which mines of gold and silver are discovered, and the medicinal plants [to be found in the wild] ... After a long discourse on the emptiness of human greatness, he drew from a leathern pouch a very few small pearls, which he forced us to accept ...'

They also met a man who suckled his child, observed a solar eclipse, even counted the different types of lice to be found in the women's hair. Humboldt discovered his first new plant species, a shrub he called *Avicennia tomentoso*. But late in October a club-wielding Indian madman on the beach at Cumaná attacked Humboldt and Bonpland; Bonpland received a serious concussion and both men decided it was time to leave Cumaná.

'Men, naked to the waist'

In November of 1799 Humboldt and Bonpland moved their base of operations farther west to Caracas, the capital of Venezuela. Humboldt's plan was daring in the extreme. He and Bonpland would travel across the Llanos, the vast and arid central plains of Venezuela, to the Orinoco River, which they would follow to its tributary the Atabapo River. But they also wanted to investigate a river called the Casiquiare, which reportedly connected the Orinoco and the Amazon—if so, it would be a rare time in nature when a single river provided water for two huge river systems. (The Casiquiare, the largest river on earth to connect two river systems, flows into the Negro River, which in turn flows into the Amazon.)

In March 1800 Humboldt and Bonpland journeyed to the Llanos. Along the way, they stopped at Lake Valencia, whose water levels had shrunk drastically in the last few years. The locals were convinced that a huge underwater cavern was siphoning the water away, causing what some feared would be 'the total disappearance of the lake'. Humboldt advised planters to sink granite pillars into the lake, in order to judge exactly how much water was disappearing, but did not believe any stories about secret caverns. His explanation was one that modern scientists and environmentalists now embrace:

> *The changes, which the destruction of the forests the clearing of plains, and the cultivation of indigo have produced within half a century in the quality of water flowing in, on the one hand, and, on the other,*

the evaporation of the soil and the dryness of the atmosphere present causes sufficiently powerful to explain the progressive diminution of the lake of Valencia ... When forests are destroyed, as they are everywhere in America by the European planters, with an imprudent precipitancy, the springs are entirely dried up, or become less abundant.

Travelling much of the time at night to avoid the heat of the sun, they reached the Llanos. The earliest conquistadors arriving in these regions had found themselves lost on the endless plains, which Humboldt found awesome in their size and loneliness, comparing them to a vast ocean: 'the resemblance to the surface of the sea strikes the imagination most powerfully'. He and Bonpland felt 'the uniform landscape of the Llanos; the extremely small number of their inhabitants; the fatigue of traveling beneath a burning sky, and an atmosphere darkened by dust; the view of that horizon, which seems for ever to fly before us; those lonely trunks of palm-trees, which have all the same aspect, and which we despair of reaching ...'

They encountered South American cowboys they considered to be nearly as wild as the area itself:

... men, naked to the waist and armed with a lance, ride over the savannahs to inspect the animals; bringing back those that wander too far from the pastures of the farm, and branding all that do not already bear the mark of their proprietor. These mulattos, who are known by the name of peones llaneros, *are partly freed-men and partly slaves. They are constantly exposed to the burning heat of the tropical sun. Their food is meat, dried in the air, and a little salted; and of this even their horses sometimes partake. Being always in the saddle, they fancy they cannot make the slightest excursion on foot ...*

Despite their astonishment at the scale and size of the Llanos, the two travellers were happy to see them end—the temperatures reached 41 degrees Celsius (106 degrees Fahrenheit) daily and the land around them, as they travelled, was filled with mirages—two suns in the sky, waters, streams, trees and mountains, none of which existed.

Fishing with horses

Before Humboldt and Bonpland left the steppes, however, they stopped at the trading station at Calabozo, where they met an educated local inventor named Señor del Pozo, of whom Humboldt asked a favour. Humboldt had long been fascinated by the electric eels that thrived in the local ponds—these were yellow, 2.7 metres (9 feet) long and could deliver a charge of about 650 volts, enough to kill a human being—but on their travels he had found them too dangerous to catch. Could Señor del Pozo help them? Indeed, he could, showing the travellers an ingenious, if barbaric, approach to catching the eels.

Pozo had his cowboys drive thirty horses into a pond filled with eels and then keep the horses trapped there with their long lances. Humboldt watched the scene unfold with astonishment:

> *These yellowish and livid eels, resembling large aquatic serpents, swim on the surface of the water and crowd under the bellies of the horses ... Several horses sink beneath the violence of the invisible strokes, which they receive from all sides in organs the most essential to life, and, stunned by the force and frequency of the shocks, disappear under the water. Others, panting with mane erect and haggard eyes expressing anguish, raise themselves and try to flee from the storm by which they are overtaken.*

But the cowboys drove them back. Within ten minutes, however, the eels had run out of charges and Humboldt was able to have five pulled in with harpoons. They were not harmless, however. 'I very stupidly placed both feet on an electric eel that had just been taken out of the water,' Humboldt wrote. 'I was affected for the rest of the day with a violent pain in the knees and almost every joint.' Disappointingly he could not get the eels to register on his electrometer.

The man who gave milk

When Alexander von Humboldt arrived in Venezuela, he was absolutely fascinated by the story of an Indian man who 'suckled a child with his own milk'. The man, a peasant named Francisco Lozano, supposedly began to breastfeed his baby son after the child's mother became ill. Humboldt wrote:

> The father, to quiet the infant, took it into his bed, and pressed it to his bosom. Lozano, then thirty-two years of age, had never before remarked that he had milk: but the irritation of the nipple, sucked by the child, caused the accumulation of that liquid. The milk was thick and very sweet. The father, astonished at the increased size of his breast, suckled his child two or three times a day during five months. He drew on himself the attention of his neighbors, but he never thought, as he probably would have done in Europe, that anything untoward occured ...

Humboldt and Bonpland went to the village where the man lived and, when it turned out Lozano was not home, left an invitation for him to come to their quarters. This he did, and bared his breasts: 'M. Bonpland examined with attention the father's breasts, and found them wrinkled like those of a woman who has given suck. He observed that the left breast in particular was much enlarged; which Lozano explained to us from the circumstance, that the two breasts did not furnish milk in the same abundance.'

Humboldt goes on to spend nearly five pages speculating on this phenomenon and his report caused a stir among his European readers. Much of his speculation is tinged with the prejudices of the day—'the anatomists of St Petersburgh have observed that, among the lower orders of the people in Russia, milk in the breasts of men is much more frequent than among the more southern nations'—but male lactation is relatively common. Men have mammary glands and, whether through hormonal changes due to drugs (often used to treat prostate cancer) or through intense sucking stimulus, as apparently happened with a dedicated father like Lozano, they can produce milk.

Orinoco journey

Humboldt and Bonpland finally reached the Apure, where they hired an Indian crew and travelled in dugout canoes down to the Orinoco. After the arid savannah of the Llanos, the river and jungle teemed with life. Humboldt's descriptions of the natural scene are typically lyrical:

> *Everything passed tranquilly till eleven at night; and then a noise so terrific arose in the neighbouring forest, that it was almost impossible to close our eyes ... These were the little soft cries of the sapajous, the moans of the alouate apes, the howlings of the jaguar and couguar, the peccary, and the sloth, and the cries of the curassao, the parraka, and other gallinaceous birds. When the jaguars approached the skirt of the forest, our dog, which till then had never ceased barking, began to howl and seek for shelter beneath our hammocks.*

The travellers arrived at the Orinoco in early April, and were astonished by its size. 'An immense plain of water stretched before us like a lake,' Humboldt wrote. 'White topped waves rose to the height of several feet.' Like the Llanos, the Orinoco produced a hallucinatory effect: 'A vast beach, constantly parched by the heat of the sun ... resembled at a distance, from the effect of mirage, pools of stagnant water. These sandy shores, far from fixing the limits of the river, render them uncertain by enlarging or contracting them alternately, according to the variable action of the solar rays.'

Humboldt and Bonpland embarked on this vast body of water in a single, 12-metre (40-foot) long dugout canoe. They spent the next month on the Orinoco, 'one of the most majestic rivers of the New World'. They were entering an area where few Europeans had ever trod. 'The crocodiles and boas are masters of the river,' Humboldt marvelled. 'The jaguar, the peccary, the tapir, and the monkeys traverse the forest without fear and without danger; there they dwell as in an ancient inheritance.'

After taking the Orinoco to the Negro River, they canoed down it and, on 10 May, reached the Casiquiare. Journeying back towards the Orinoco on the Casiquiare, the two men were plagued by the excessively humid weather, which caused Bonpland's plant specimens to rot, and by fierce mosquitoes, and were somewhat dismayed to find that even their campfires did not deter the ever-present jaguars from coming close to camp (although Humboldt reassured himself that there was only one case he knew of when a person had been attacked in his hammock). He and Bonpland found out about another danger—what they called the 'caribe' fish, otherwise known as piranhas. To test how dangerous these 10-centimetre (4-inch) long fish really were, they tossed raw meat into the water: 'In a few minutes, a perfect cloud of caribes came to dispute their prey.' The men, travelling in a narrow and unstable canoe, were suitably chastened.

On 22 May 1800, Humboldt and Bonpland reached the Orinoco, thus proving that the Casiquiare did indeed feed both river systems, an achievement Humboldt considered was the most important he would make during his South American journey. Travelling back downstream, they reached the town of Angostura (from whence the world-famous bitters come), where they both caught typhoid. It took them a month to become strong enough to even consider crossing the Llanos again, but they finally managed to so do, returning to Caracas and then taking ship back to Cumaná. From there they booked passage to Havana, arriving in December, eighteen months after the *Pizarro* had first brought them to the New World.

'How the forces of nature interact'
The two men had made a 2500-kilometre (1500-mile) journey through wilderness unseen by most Europeans and returned with twelve thousand specimens of plants, insects and animals. Theirs had been an extraordinary success. Not only had Humboldt and Bonpland proved that the Casiquiare joined the Amazon and Orinoco, they had also studied the profusion of wilderness in South America in

a way never studied before. Before going on his voyage, Humboldt had written to his brother: 'I shall try to find out how the forces of nature interact upon one another ... I must find out about the unity of nature.' Humboldt comes across as a modern scientist; he understood that forces within nature intertwined and reacted with one another, that cutting down trees affects the depth of lakes, that food chains exist. He was awed by the primitive wilderness they had travelled through because 'we almost accustomed ourselves to regard men as not being essential'. Instead of being the centre of Creation, humans were merely another organism, and in the vast Llanos and the teeming jungles of the Orinoco and Casiquiare not even a very necessary one.

Despite their exhausting journey, Humboldt and Bonpland were not through with South America. After sending home their specimens and copies of their notes and journals, they heard that Bougainville's around the world expedition was on again. Hoping that they could meet it in Lima, Peru, Humboldt and Bonpland made another extraordinary journey from Havana to Quito, Ecuador. Along the way they passed through the high Andes, mostly on treacherous paths with sheer drops along the side. They climbed 6000-metre (19,000-foot) tall Chimborazo, nearly making it to the summit before they were stopped by an unbridgeable chasm, but even so climbing higher than any human being on earth at the time.

In Quito they heard that the Bougainville expedition was not going to touch in South America. Undaunted, Humboldt and Bonpland continued on to Lima, descending from the mountains on the fine, ancient Incan roads and reaching Lima in October of 1822. They brought with them, according to Humboldt, as many as twenty mules loaded with specimens. After spending two months in Lima, they embarked for Mexico City and to North America, where their fame had become so widespread that they were able to dine with President Thomas Jefferson himself. But Humboldt had already gone through a good deal of his fortune and decided it was time for them to head back to Europe, where they arrived in August of 1804, to be lionised by the notables of the Continent (if not by Napoleon).

This was to be their last grand journey together. Aimé Bonpland went on to work for the Empress Josephine in Paris, while Humboldt returned to Prussia. He spent the rest of his life, while his fortune slowly dwindled, in scientific pursuits that included studying electrical storms and writing the thirty-five volume story of his journeys with Bonpland. By the time he died at the age of ninety in 1859—the same year that Charles Darwin published *The Origin of Species by Means of Natural Selection*—Humboldt was dependent on a small government stipend to make ends meet and left everything he had to an ageing servant. Many young scientists no longer knew who he was. As author Stephen Bown has written of him: 'His influence on European science was so great that by the time of his death, many of his theories were considered standard subjects of study or were so widely accepted that his contributions were unrecognized ... It was probably either his greatest source of pride or his greatest disappointment.'

CHARLES WATERTON
The squire of Walton Hall

The English naturalist Charles Waterton would often startle guests to his Yorkshire manor, Walton Hall, by hiding behind the front door and biting them as they entered. (His pretensions to being a dog were no idle fancy—he was perfectly capable, even when well past sixty, of scratching the back of an ear with his big toe.) Presumably any guest to Walton Hall would forgive their host for taking a chomp or two out of their ankles, because Waterton was, in fact, a very entertaining fellow. After all, how many people do you meet who have knocked out a boa constrictor with just one punch? Or tried to fly from the top of their own outhouses ('navigating the atmosphere', as Waterton called it)? The twenty-seventh squire of Walton Hall loved to climb, scampering up trees well into his eighth decade. Even more startling, Waterton, a staunch if eccentric Roman Catholic, once climbed to the top of St Peter's Basilica in Rome, where he deposited one of his gloves on a lightning rod as a kind of calling card. And when the pope asked him to please remove it, he climbed back up and did so.

Of course, the nineteenth century abounded with dotty English aristocrats—a contemporary of Waterton's, Squire John 'Mad Mytton' Mytton, once disguised himself as a robber and shot two of his own guests—but the difference in Waterton's case is twofold. One, he was one of the most brilliant field naturalists the early nineteenth century had seen. And, two, he had a gentle heart and a generous soul (unless you differed from him on some fine point of bird lore). How can you not love a man who penned this self-description, when aged fifty:

> *I stand six feet, all but a half an inch [1.81 metres]. On looking at myself in the glass, I can see that my face is anything but comely: continual exposure to the sun, and to the rain of the tropics, has furrowed it in places, and given it a tint which neither Rowland's Kalydor, nor all the cosmetics of Belinda's toilette, would ever be able to remove. My hair, which I wear very short, was once of a shade twixt brown and black: it now has the appearance as though it had passed the night exposed to a November hoar frost. I can-not boast any great strength of arm; but my legs, probably by much walking, and by frequently ascending trees, have acquired vast muscular power ... To speak zoologically, were I exhibited for show at a horse fair, some learned jockey would exclaim, he is half Rosinante, half Bucephalus.*

'The dogs howled fearfully'

This cross between Don Quixote and Alexander the Great (or at least, their trusty steeds) was born in Walton Hall, near Wakefield, Yorkshire, in 1782. He was from a family of aristocratic Roman Catholics who had survived the Reformation while remaining staunch supporters of the English crown. When he was ten years old Waterton was sent to a Catholic school (where he bit the leg of a priest who was flogging him, the beginning of a lifetime of leg biting) and from there made his way to Stonyhurst, where he studied Latin but really excelled at natural history. Waterton later wrote of his school years:

> *I was considered rat-catcher to the establishment, and also fox-taker, foumart-killer, and cross-bow charger at the time when the young rooks were fledged ... I followed up my calling with great success. The vermin*

disappeared by the dozen; the books were moderately well-thumbed; and according to my notion of things, all went on perfectly right.

Even then Waterton was eccentric—from an early age he wore his hair short, almost in what we today would call a crew cut, which was unfashionable during an era of long hair—and upon his graduation his parents sent him to Malaga, in Spain, with a couple of his maternal uncles, possibly to season him. He got more seasoning than anyone bargained for when Malaga was hit by the pneumonic plague—the 'black vomit', as it was called—which killed both of his uncles. It was a terrible scene, Waterton wrote in his memoirs: 'The dogs howled fearfully during the night. All was gloom and horror on every street; and you might see the vultures on the strand, tugging at the bodies which were washed ashore by the eastern wind.'

Malaga was quarantined; nonetheless, Waterton fled back to England, where, as he wrote in his memoirs: 'I was seized with vomiting and fever during the night. I had the most dreadful spasms and it was supposed I could not last out the day. However, strength of constitution got me through.'

Strong constitution or not, what Waterton called 'the bleak and wintry wind of England' drove him to warmer climes after he had recuperated. In 1804 he sailed for Demarara, the British city in Guyana, on the northern coast of South America. One of his paternal uncles had an estate there and his father had recently purchased one, and so Waterton was given what was, to him, the happy task of managing these properties. The Napoleonic wars were 'becoming general', as he later wrote, and the tropics now seemed particularly inviting.

'Easy prey to any alligator'

Before heading to South America, Waterton was introduced to Sir Joseph Banks (see page 79), who gave him advice on how to survive in the tropics. 'You may stay in them,' he told young Waterton, 'for three years or so and not suffer much. After that period, fever, ague, and probably a liver disease, will attack you, and you will die at last, worn out, unless you remove in time to a more favourable climate.'

Banks went on that since Waterton 'did not have his bread to seek' (he did not need to make a living), he should come home after three years or so, and thus beat the odds of dying of tropical fever. Waterton did not take this advice to heart. He lived in what was then Dutch Guyana for eight years (Great Britain assumed control in 1815, and it became known as British Guyana), looking after his father's and his uncle's sugar plantations. He had not yet begun his expeditions as a naturalist, but he did make journeys into the interior. On the first of these, in 1809, he made his way up the Sacopan River: 'a grand feast for the eyes and ears of the ornithologist' as 'water fowl innumerable', including myriad multi-coloured parrots, flew overhead.

During this journey, there occurred an incident that was typical of the misadventures Waterton would experience. As the boat passed by the riverbank, Waterton noticed a *labarri* snake—the deadly poisonous fer-de-lance—coiled in a bush. As he tells the story:

> *I fired at it and wounded it so severely that it could not escape. Being wishful to dissect it, I reached over into the bush, with the intention to seize it by the throat and convey it onboard. The Spaniard at the tiller, on seeing this, took the alarm and immediately put his helm aport. This forced the vessel's head to the stream, and I was left hanging to the bush with the snake close to me, not having been able to recover my balance when the vessel veered from land. I kept firm hold of the branch to which I was clinging, and was three times over-head in the water below, presenting easy prey to any alligator that might have been on the look out for a meal.*

Finally the crew pulled him in, soaking wet, and he insisted on going back for the snake, much to their great discomfort. The animal measured 2.5 metres (8 feet) in length, and Waterton happily dissected it.

'The native country of the Sloth'

By 1812 Waterton's father and uncle had both died and he returned to England to claim his large inheritance. He then returned to Guyana, promptly handed over management of the plantations to others and embarked on the series of journeys into the South American jungle that he immortalised in his classic book *Wanderings in South America*, which covers four separate journeys. The first trip, begun in April 1812, was in search of curare, the fabled poison used by the Macusi Indian tribes, among others, to coat their arrows. It renders even a slight wound to their prey (either animal or human) a deadly one by paralysing the respiratory muscles. Waterton sought out such a deadly substance because he had read medical research that postulated curare might work as a possible antidote to hydrophobia and tetanus, since there was some evidence that small doses acted as a muscle relaxant.

This would be a round-trip journey of more than 1500 kilometres (1000 miles) and would take Waterton through jungle mainly unexplored by Europeans and as far as the Portuguese frontier fort of San Joachim on the Rio Branco, a tributary of the Amazon. Setting off, quite alone, Waterton soon passed through the last vestiges of civilisation and plunged into the jungle, following the winding path of the Demerara River. The density of the forest might have frightened a lesser man, but it fascinated Waterton. The trees, he observed, were not thick in girth, but they made up for that in their height and their beauty and usefulness. Waterton lists them: 'The Green-heart, famous for its hardness and durability; the Hackea, for its toughness; the Ebony and Letter-wood, vying with the choicest woods of the old world; the Locust-tree, yielding copal; and the Hayawa and Olou-trees, furnishing a sweet smelling resin ...'

One gets the sense of Waterton's being overwhelmed by the beauty and grandeur around him as birds squawked in the trees, monkeys howled and herds of wild pigs rustled through the forest. Waterton, who loved wild sloths (these eccentric creatures were every bit as odd as he was), watched them carefully on this journey. 'This ...

is the native country of the Sloth,' Waterton wrote. 'His looks, his gestures, and his cries, all conspire to entreat you to take pity on him ... While other animals assemble in herds, or in pairs range through these boundless wilds, the sloth is solitary, and almost stationary; he cannot escape from you. It is said, his piteous moans make the tiger relent, and turn out of the way. Do not level your gun at him, or pierce with a poisoned arrow—he has never hurt one living creature.'

Wouralia, the ass

Charles Waterton was fascinated by curare and went to great lengths to acquire some. The Macusi Indians either allowed him to be present during the ceremonies that involved making the poisonous substance or told him some of the details, as he was able to list with some accuracy numerous of the plant ingredients that went into curare (chief among them the wourali vine, or Strychnos toxifera*) and relate many of the taboos surrounding its preparation.*

When Waterton returned to his home in Guyana, he immediately set about testing the substance. Despite his avowed love of sloths, a sloth was one of the first creatures to feel the sting of Waterton's curare:

> The [sloth] was wounded in the leg and put down on the floor, about two feet [60 centimetres] from the table ... It sank to the ground but sank so gently, that you could not distinguish the movement from ordinary motion. During the tenth minute from the time it was wounded it stirred and that was all; and the minute after, life's last spark went out.

Next on the list, while Waterton was still in Guyana, was a 450-kilogram (1000-pound) ox, shot with three curare-tipped arrows. Its death within a few minutes proved to Waterton that the poison could fell larger animals.

But it was back at Walton Hall that Waterton conducted his greatest experiment, on a young donkey he nicknamed Wouralia. In a scene that has unavoidable overtones of a mad scientist movie, Waterton injected the ass with poison and it died (apparently) in the usual ten minutes. But:

An incision was then made in its windpipe, and through it the lungs were regularly inflated for two hours with a pair of bellows. Suspended animation returned. The ass held up her head, and looked around; but the inflating being discontinued, she sunk once more in apparent death. The artificial breathing was immediately recommenced, and continued without intermission for two hours. This saved the ass from final dissolution; she rose up and walked about; she seemed neither in agitation or in pain ...

Perhaps sensing that this experiment was beyond the pale, even for him, Waterton is at pains to assure the reader that Wouralia, after recuperating, lived another 'five and twenty years'. His experiment was a precursor to another in the twentieth century that showed that curare can be used, along with general anaesthesia, to relax chest muscles during abdominal surgery.

The White Sea

Waterton stumbled upon Indian huts in a cleared area in the woods about two weeks into his journey, near where the Demerara fell in great torrents down a chasm. He made it clear to the Indian that he was looking for curare—*wourali*, as Waterton called it, using the local term. He had to pay a very high price for it, but he understood— '[curare] was powder and shot to them, and very difficult to be procured'. He tested the powder on a dog, merely placing a tiny amount into a wound on the animal's thigh. Within fifteen minutes, the creature was dead.

After obtaining his curare, he engaged some Indian guides and they took him through the forest. About two hours below the falls, at midnight on 1 May, Waterton heard strange noises from the forest, 'as though several regiments were engaged, and musketry firing with great rapidity'. The Indians with him were terrified and they all huddled together until dawn, but no reason for the disturbance was ever found, an experience that foreshadowed that of Henry Bates when he heard strange noises in the Amazonian rainforest (see page 213) years later.

After a few more weeks, Waterton entered the territory of the Macusi Indians, a people who were 'uncommonly dextrous in the use of the blow-pipe, and famous for the skill in preparing the deadly vegetable poison, commonly called Wourali'. Waterton had sought out the Macusi because he understood that they made a particularly powerful brand of the poison—it was from them he sought to procure the greatest amount. Unfortunately, they too were stingy with their curare, although he finally managed to purchase a small supply.

While he was in the village of the Macusi, the Indians told him that four days march away was the fabled Lake Parima, or White Sea, where supposedly the king of the mythical city of El Dorado took his baths after coating himself with gold dust. Waterton did not believe the story—nor did he believe that a race of Indians with tails lived there, as the Macusi told him—but his curiosity got the better of him and he decided to journey there. After a long trek through jungle and across a broad savannah ('the finest park that England boasts falls far short of this delightful scene') the Indians could show him only a 'spacious plain' that, when covered with rainwater, had 'the appearance of a lake'. Waterton speculated that the plain may once have been the bed of a huge lake—something that modern geologists have confirmed.

'To have been once sucked by a vampire'

Waterton then proceeded through the jungle to the Portuguese frontier fort of San Joachim, first sending along a note to its commander apprising him that he was coming. The commander wrote back politely, saying that his government forbade him to allow strangers to enter the fort but indicating that he would meet Waterton a short distance from San Joachim. By the time Waterton arrived there, ferried by the Macusi, he was shivering and ill with fever, probably malaria, and the commander took pity on him and took him into the fort.

Waterton is not clear quite how long he stayed at the fort, recuperating, but it was probably several weeks. Afterwards, still weak from fever, he journeyed back to the coast where he rested at the home of friends and then returned to England and Walton Hall. There he remained for three years as recurring bouts of malaria made it inadvisable for him to travel. However, he did not waste his time merely lying about. Using the curare he brought with him, he conducted a series of bizarre animal experiments that did, however, demonstrate the usefulness of curare for medical purposes.

With his health finally restored, Waterton set off back to South America on his second journey, arriving in Guyana in March of 1816. In his book, before describing his arrival there, he gave his readers some idea of what to take with them, should they decide to visit 'the torrid zone'. Flannel underwear was a must ('a great preserver of health in the hot countries') but, after that, 'a hat, a shirt, and a light pair of trousers will be all the rainment you require'. However, 'custom will soon teach you to tread lightly and barefoot on all the little inequalities of the ground, and show you how to pass on, unwounded, amid the mantling briars'.

Waterton also advised readers to carry with them on their journeys three drugs—laudanum, calomel and jalap—and to bring a lancet for bloodletting. Waterton's own theory for surviving in the tropics (aside from not spending too much time there) was based on 'abstemiousness', especially in regard to hard liquor or opium, so he advised cautious use of the laudanum. Calomel is a laxative (it was also used for treatment of syphilis), while jalap is a powerful natural purgative. As for the lancet? Well, Waterton was a firm believer in bleeding himself for his health. ('I now opened a vein,' he writes on one occasion when he was ill in the jungle, 'and made a large orifice, to allow the blood to rush out rapidly; I closed it after losing sixteen ounces [450 grams].' He then placed his feet in a pail of warm water, dosed himself with both calomel and jalap, and innumerable cups of 'weak, warm tea', and claimed that he felt much better.)

Waterton took his love of being bled to great lengths. Writing about the vampire bats common in South America, he said:

> *I often wished to have been once sucked by a vampire, in order that I might have it in my power to say it had really happened to me. There can be no pain in the operation, for the patient is always asleep when the vampire is sucking him; and as for the loss of a few ounces [50 grams or so] of blood, that would be a trifle in the long run. Many a night have I slept with my foot out of the hammock to tempt this winged surgeon; but it was all in vain, the vampire never sucked me ...*

'Guiana still whispered in my ear'

On his second journey, Waterton did not venture very deep into the interior but stuck to coastal areas in Guyana and Brazil. On this journey, he focused to a great extent on birds and, in fact, after Humboldt, was probably one of the first European naturalists to observe South American birds closely. His descriptions of hummingbirds, parrots of all types, large-billed toucans, bitterns, egrets and blue herons are lyrical. He also dispelled a myth concerning the goat-sucker, a migratory, night-flying bird that spends a great deal of time hovering around the bellies of goats and cows. The Indians thought that the goat-sucker sucked milk from their animals, but close observation by Waterton showed otherwise—capturing one of the animals and dissecting it, he discovered that the bird's stomach was filled with flies that would otherwise annoy the animal herds.

After spending nearly a year in the jungle, Waterton once again returned to England, staying there for three years, with a side trip to Italy (where he deposited his glove atop St Peter's). But, as he wrote, 'Guiana still whispered in my ear', and he decided to set forth again, this time landing back in South America in February 1820. Using the home of his friend John Edmonstone, whose daughter Mary Anne he would later marry, as a base for operations, he set forth again.

It was deep in the jungle on this third journey that Waterton had his famous bout of fisticuffs with a boa constrictor. Walking through the jungle one day, he observed a young boa about 3 metres (10 feet) long moving slowly along a trail. Most people would take this as an excellent excuse to turn around and walk in the other direction but, of course, Charles Waterton did not.

> *I saw the snake was not thick enough to break my arm in case he got twisted round it. There was not a moment to be lost. I laid hold of his tail with the left hand, one knee being on the ground; with the right I took my hat, and held it as you would a shield for defence. The snake immediately turned and came at me, with his head about a yard [a metre] from the ground, as if to ask me, what business I had to take liberties with his tail. I let him come, hissing and open-mouthed, within two feet [60 centimetres] of my face, and then, with all the force I was master of, drove my fist, shielded by my hat, full in his jaws.*

The snake was, naturally, 'stunned and confounded' by the blow, which gave Waterton time to wrap his hands around the poor reptile's neck and carry it back to camp, where he killed it. He then left the carcase out in the forest and, sure enough, he was soon treated to the sight of the bird that Waterton called the 'King of Vultures', actually the king vulture, an extremely powerful vulture found in Central and South America. After describing the creature's majestic appearance, along with its characteristic blue and orange nose carbuncle, Waterton even closely depicts 'the bag of the stomach, which is only seen when distended with food, [and] is of a most delicate white intersected with blue veins, which appear on it just like the blue veins on the arm of a fair-complexioned person'.

It was also on this eventful third journey that Charles Waterton had his fabled bareback ride on a cayman, before heading down a fierce set of rapids to the ocean. He then travelled back to Guyana, from whence he returned to England in 1821.

Riding the beast

Charles Waterton had long wanted to capture a specimen of the animal he called a 'cayman' (which may have been either an alligator or a crocodile). One day on his third journey to South America, he was told by some Indians that a 3-metre (10-foot) long specimen of the creature lived nearby. He paid a few of the Indians to help him trap it with several hooked arrows tied to the end of a rope, but when they offered to shoot the creature for him, either with arrows or guns, he did not want to mutilate this fine specimen.

In classic Waterton fashion, he settled on another solution. His canoe had a mast, about 2.4 metres (8 feet) in length, with a sail attached. Waterton ordered that the mast be brought to him and he had the sail wrapped around one end of it. He then approached the cayman, where it was being held by ropes, intending to ram the timber down the reptile's throat. Getting close to the creature, Waterton wrote with hilarious understatement, 'I saw enough not to fall in love at first sight'. But when the animal was dragged to within a few metres of Waterton, he changed his mind and decided that he would go for a ride, instead. As he tells it:

> I instantly dropped the mast, sprang up, and jumped on his back, turning half round as I vaulted, so that I gained my seat with my face in the right positions. I immediately seized his forelegs and, by main force, twisted them on his back; thus they served me as a bridle.

The Indians, still holding onto the cayman by means of ropes, dragged the animal and its rider 35 metres (40 yards) or so up the riverbank, cheering the whole time. The cayman finally exhausted itself and lay still, and Waterton was able to tie its jaws shut and cut its throat. An illustration of the episode in Wanderings in South America *became famous, and it was one of Waterton's more celebrated episodes, though, of course, there was absolutely no scientific reason for riding on the back of the beast.*

'In Equatorial regions'

Waterton's fourth and last journey, starting in 1824, began in North America, where he visited the northeast, wandering through upstate New York in time to see the epic deforestation of that region— 'Most of the stately timber has been carried away,' Waterton wrote, 'thousands of trees are lying prostrate on the ground ... I wish I could say a word or two for the fine timber which is yet standing. Spare it, gentle inhabitants, for your country's sake.' But in all he enjoyed America, except for the bad roads, on one of which he badly sprained an ankle. He was astonished at the spectacle of Niagara Falls, and also by the 'unaffected ease and elegance of the American ladies' he saw there. Ever the gentleman, Waterton wrote that the more he looked at American women, 'the more I was convinced, that in the United States of America may be found grace and beauty and symmetry equal to anything in the old world'.

After visiting Canada, Waterton set sail for Antigua, and finally back to Guyana. Now forty-two years old, he realised that this would be his last time visiting his beloved jungles and, like a fond lover saying good-bye to old haunts, he revisited his favourite animal, the sloth, keeping one in his house with him for several days. 'On the ground,' Waterton wrote with exasperated affection, '[the Sloth] appeared really a bungled composition, and faulty at all points; awkwardness and misery were depicted on his countenance; and when I made him advance he sighed as though in pain'.

Waterton still wanted deeply to have his toe sucked by a vampire bat. During this last visit, 'a young man of the Indian breed', sleeping in his hammock next to Waterton, had his big toe sucked in this fashion. Waterton, beside himself with jealousy, examined it after the young man had bathed it in the river the next morning: 'The midnight surgeon had made a hole in it, almost of a triangular shape, and the blood was running from it apace.' To add insult to injury, an old woman whom Waterton paid to wash the young man's bloody hammock told him that vampire bats 'generally preferred younger people'.

Returning to England in early 1825, Waterton finished the last chapter of *Wanderings in South America* and sent it to his publisher. It closed with a short verse he had written:

> *And who knows how soon, complaining*
> *Of a cold and wifeless home,*
> *He may leave it, and again in*
> *Equatorial regions roam?*

The squire of Walton Hall

Waterton's book proved to be very popular, going through numerous editions and remaining in print until the first decade of the twentieth century. It was read by the young Charles Darwin and other naturalists of the day, who thrilled to Waterton's encounters and his descriptions of animal and plant life.

For a time, it seemed Waterton's life would be a happy one. In 1829, at the age of forty-seven, he married Mary Anne Edmonstone, John Edmonstone's seventeen-year-old daughter, and found himself quite enraptured by happiness. Unfortunately, the following year she died giving birth to his son, Edmund, and Waterton—who blamed himself for not being able to save her from death—grieved mightily. He began sleeping on the bare wooden floor with only a block of wood for a pillow, and it was now that he began biting his dinner guests.

Walton was to live for another thirty-five years, dying in 1865 after falling over a log on his estate but retaining his physical fitness right up until the end. One of his friends said that, at the age of seventy-nine, he could still, with a head start, leap over a 1.2-metre (4-foot) high wire fence without touching it. But Waterton was not merely engaged in eccentricity and fence-hopping, for at Walton Hall he started the very first nature preserve the world had ever seen. He enclosed 100 hectares (250 acres) of his estate with a 2.4-metre (8-foot) high fence that ran for 5 kilometres (3 miles) and turned it into a bird sanctuary, fighting to keep out poachers. He hated indiscriminate killing of wildlife; when a neighbour's gamekeeper handed him a bittern he had just shot, expecting thanks, Waterton sneered: 'Your father murdered the last raven in England', and walked away. He hated, as well, the encroaching factories of the Industrial Revolution, with their 'volumes of Stygian smoke' and 'filthy drainage [polluting] the waters in every river far or near'.

It was perhaps his last stance as an early ecologist, well ahead of his time, that made Charles Waterton seem especially eccentric to those around him. To us, he merely seems prescient.

The fate of Walton Hall

Charles Waterton's only son, Edmund, did not keep up the property and nature preserve at Walton Hall as his father had hoped. After Charles's death, in fact, Edmund, in need of money, held shooting parties in the preserve, killing the birds that his father had tried so zealously to protect.

To make matters worse, Edmund finally sold the estate, after fourteen generations of Watertons had lived there. Its purchaser was Edward Simpson, son of an old enemy of Charles, the chemical and soap works owner 'Soapy' Simpson, whose polluting factories were exactly the ones that the naturalist so reviled. The Simpsons lived in Walton Hall until 1940. During World War II, Walton Hall was pressed into service as a military hospital and after the war it became a maternity hospital, but it fell into disuse and disrepair in the 1970s and was almost torn down.

However, in the 1990s it was restored and turned into the Waterton Hotel conference hall and meeting centre. The old reserve is a golf course, while the stables and farm buildings have been converted to private residences.

PRINCE MAXIMILIAN OF WIED
'a solemn silence prevailed'

While sailing across the Atlantic Ocean at the start of his journey to the United States in the spring of 1832, Prince Maximilian of Wied noted in his journal that 'voyages to America are become everyday occurrences and little more is to be related of them than that you met and saluted ships, had fine or stormy weather, and the like'. Actually, such voyages could be hair-raising adventures in and of themselves, despite the prince's insouciant attitude; however, he did entertain readers of his memoirs with one titbit. He described sailors harpooning dolphins ('monsters of the deep', the prince called them) and then for dinner serving up dolphin steaks, which the prince and his companions 'found to be very good; indeed, we preferred them to all other meat'. And then Maximilian went on: 'I did not know at the time that I should soon find a dog's flesh relishing.'

Indeed he would. For the forty-nine-year-old Alexander Philipp Maximilian—ruler of the Prussian state of Wied-Neuwied—was embarked on one of the earliest scientific journeys through the American West ever taken, a two-year trek up the turbulent Missouri River and into the Rocky Mountains. It would be extraordinarily arduous but would reward the prince with a unique glimpse of a wilderness where few Europeans had gone before, a land where, as he wrote, 'a solemn silence prevailed ... where Nature, in all her savage grandeur, reigned supreme'.

'Cheerless, desolate prairies'

Maximilian was born in Wied-Neuwied (now a part of Germany) in 1782. He was raised in a castle on the Rhine, the eighth of ten children and the second son of Prince Johann Friedrich Alexander, ruler of Wied-Neuwied. At an early age he became a student of Johann Friedrich Blumenbach, an important scholar in the new field of physical anthropology. However, the Napoleonic Wars interrupted Maximilian's studies. After entering military service, he was captured by the French during Napoleon's decisive victory over the Prussians at Jena in 1806. After he was released he became a major general and was to lead Prussian troops into Paris at the war's end in 1814.

After the war, Maximilian returned to naturalism. He had met Alexander von Humboldt (see page 95) while he was in Paris and the latter became a mentor to the young Prussian prince, encouraging him to journey to South America to study the flora, fauna and people there. Maximilian lived in Brazil for two years; then he returned to Europe and over the course of the next decade compiled a book about his experiences *(Travels in Brazil in the Years 1815, 1816, 1817)*. Possibly because of his studies with Blumenbach, Maximilian was especially interested in the Indians of South America and spent a great deal of time observing them. When *Travels in Brazil* was published it was translated into four languages and brought the prince a good deal of popular and scholarly acclaim. Prompted by this success, he decided to travel to North America. As he told readers in his book *Travels in the Interior of North America*, he wanted to provide 'a clear and vivid description of the natural scenery of North America' beyond its current frontiers, scenery that included 'those cheerless, desolate prairies, the western boundary of which is formed by the snow-covered chain of the Rocky Mountains'. The United States, he further wrote, 'may be justly reproached for not having done more to explore them'.

'The original American race'

In fact, almost thirty years before, the United States had mounted a massive expedition up the Missouri River to the Pacific coast—the famous Lewis and Clark expedition, the first to explore the region. But there had been little in the way of scientific inquiry into the vastness west of the Mississippi River—and with good reason. The American West was a daunting unknown. The Missouri River rises in the Rocky Mountains and travels east to the Mississippi, covering some 4000 kilometres (2500 miles), making it the longest river in the United States. In the early nineteenth century, it was a wild, turbulent expanse, treacherous to navigate, its shoals and sandbars constantly shifting, its silt-filled waters overflowing its banks in the spring to flood the surrounding regions.

Along the banks of the Missouri (which passes through ten current-day American states) lived numerous Indian nations—the Omaha, Dakota, Crow, Mandan, Cree, Assiniboine, Blackfoot, Sioux and others. These were indeed the 'wild aborigines' of America, as Maximilian called them in his journals, and many had a nasty habit of killing whites, especially those they found alone. Wandering mountain men and trappers took their lives in their hands to bring back furs to the proprietors of the isolated fur trading posts set up by the American Fur Company, based in the booming city of St Louis, located just south of the confluence of the Missouri and Mississippi rivers.

It was into this untamed wilderness that Prince Maximilian intended to go, although not alone. He had hired a tall, handsome twenty-three-year-old Swiss artist named Karl Bodmer to sketch the natives and the local flora and fauna, taking the precaution, however, of warning him that he should not 'on the basis of my social station, make any conclusion about my standard of living since I always live very simply ... On the ship we will have good food and wine; in America this will be lacking.' Maximilian also brought with him his long-time manservant, David Dreidoppel, who had accompanied him to Brazil and who was an accomplished hunter and amateur taxidermist.

Sailing from Holland in the spring of 1832, the threesome landed in Boston on the Fourth of July, just in time to view an Independence Day celebration. Although the Americans doing the celebrating were, in the prince's words, a 'motley assemblage' and numerous booming artillery salutes shattered the peace, the Yankees were not too rowdy. However, Maximilian spoke disapprovingly of the fact that a 'stranger in Boston looks in vain for the original American race of Indians ... [America] is rapidly proceeding in the unjustifiable expulsion and extirpation of the aborigines, which began on the arrival of the European in the New World, and had unremittingly continued.'

This humanist opinion was well ahead of its time and was probably shared by few of the Americans then enjoying their Fourth of July in Boston. But if it was Indians Prince Maximilian wanted, he would soon meet his fill.

'The greasiest pair of trousers'

After spending a few days in Boston, Maximilian and his party headed to Philadelphia, where he toured a natural history museum that contained the skeleton of a mastodon (which Maximilian called an 'Ohio elephant') as well as 'most of the animals of North America, pretty well stuffed', giving him his first glimpse of the antelope, elk and grizzly bear. But an outbreak of the cholera that plagued the city in the nineteenth century hastened the prince's departure, and he headed west. He and his friends paused in New Harmony, Indiana, to visit the famous community there (as well as naturalists Thomas Say and Charles-Alexandre Lesueur), but an intended short stay turned into five months when Maximilian came down with 'a serious disposition, nearly resembling cholera'. While Bodmer made a side trip to New Orleans, Maximilian used his time to read through New Harmony's extensive library, and he also added specimens to his growing collection, which at this point consisted of five crates containing 170 birds, forty-three turtles, forty to fifty snakes, and forty frogs and toads.

Setting off on the road again in March 1833, Maximilian and his fellow travellers reached the booming town of St Louis, considered the gateway to the west. Beyond St Louis lay Indian territory, administered by Superintendent of Indian Affairs William Clark, who had become famous for his journey to the Columbia River with Meriwether Lewis. Anyone who wanted to go farther into the territory needed a 'passport' from Clark, which he was only too happy to provide Maximilian. Clark further told Maximilian that the best route into the territory was to take the American Fur Company steamer, the *Yellow Stone*, which was soon to venture up the wild Missouri to Fort Union. Maximilian had other options open to him—he could have travelled overland—but with so many hoped-for specimens to ship home, travelling by steamer seemed the best way.

Maximilian busied himself with preparations, purchasing 'coffee, sugar, brandy, candles, fine gunpowder, shot of every kind, colours, paper', and much else besides. Interestingly, the Prussian prince's appearance did not differ too widely from that of some of the other characters heading west. An American Fur Company clerk wrote this description of him:

> *In this year an interesting character in the person of Prince Maximilian from Coblenz on the Rhone [Rhine] made his first appearance ... He was a man of medium height, rather slender, sans teeth, passionately fond of his pipe, unostentatious, and speaking very broken English. His favorite dress was a white slouch hat, a black velvet coat, rather rusty from long service, and probably the greasiest pair of*

trousers that ever encased princely legs. The prince was a bachelor and a man of science and it was in this capacity that he had roamed so far from his ancestral home ...

Karl Bodmer: 'a lively, very good man'

Karl Bodmer, Maximilian of Wied wrote to a friend at the start of his journey through America, 'is a lively, very good man and companion, seems well-educated, and is very pleasant and very suitable for me; I am glad I picked him. He makes no demands and in diligence is never lacking.'

Bodmer was not lacking in one other ingredient, either—talent. His watercolours of American Indian tribes living along the Missouri River are, along with Wied's journal, the last and best record we have of Native American groups such as the Mandan, Minnataree and Akara before they perished or were decimated in the great smallpox epidemics that so reduced the native population of America.

Bodmer was born in Zurich, Switzerland, in 1809, studied art in France and thereafter worked in Switzerland as an engraver and illustrator, until he met Maximilian of Wied in 1828, on a visit to Germany. Maximilian was impressed by the young man and agreed to take him along on his journey to the American frontier—a journey that was to be the making of Bodmer's reputation. Bodmer's watercolours are vivid and accurately realistic representations of Native American customs—he captured the Bison Dance of the Mandan Indians and a battle between Blackfoot and Assiniboine Indians—but he also painted snarling grizzly bears, enormous herds of bison drinking at streams, lonely rivers and vast prairies.

Bodmer turned his watercolours into eighty-one aquatint engravings, which were reproduced in the pages of Maximilian's Travels in the Interior of North America, *an expensive book aimed at well-off Europeans. Bodmer lived out his life in France, where he painted landscapes, and died in Paris in 1893. He kept in touch with Maximilian until the prince's death in 1867. The two collaborated on an illustrated book about North American reptiles, but Bodmer was never able to replicate the brilliant realism of his Indian paintings from early America.*

'One might have been crushed'

The steamboat *Yellow Stone* set off from St Louis on 10 April 1833, to 'some guns fired' as a salute. Maximilian writes that there were 'about 100 persons on board the Yellow Stone, most of whom were called *engagés* or voyageurs, who are the lowest class of servants for the Fur Company'. These men were trappers in the employ of American Fur; there were also on board so-called 'free' trappers, independent-spirited fellows who sought their own peltries in the wilderness. Maximilian wrote that the trappers 'keep big scalping knives in a sheath on their belts. They shouted, fired their guns, and drank.'

As the *Yellow Stone* slowly progressed up the Missouri River, travelling at about 8 kilometres (5 miles) an hour against the heavy current and tying up to the bank every evening, Maximilian saw for himself how dangerous it was to navigate these muddy waters, filled with sandbars and huge, floating trees. 'The Yellow Stone had several times struck against submerged trunks of trees, but it was purposely built very strong for such dangerous voyages.' Maximilian wrote of the dense forests they passed through, dotted with isolated settlements, including one 'celebrated for the brave defense made by a few men against a numerous body of Indians'. The ship passed a huge rock jutting out of the forest—local Indian legend had it that it had been formed by the dung of a race of bison who lived in heaven. But, as Maximilian wrote with some sadness, '[the Indians] themselves no longer believed this fable'.

The *Yellow Stone* nearly ran aground in the shallow waters where the Grand River debouched into the Missouri and Maximilian was amazed at the 'devastation' caused by the spring floods; huge swamped forests of cedar seemed to sprout from the riverbed itself. On 25 April 'a large branch of a tree, lying in the water, forced its way into the cabin, carried away part of the door, and then broke off and was left on the floor'. The prince was now beginning to understand how easy it was to be killed in the wilds of America: 'One might have been crushed in bed,' he wrote. On another day, walking through the woods as the *Yellow Stone* made one of its frequent stops for firewood to stoke its boilers, Maximilian very nearly stepped on a rattlesnake, and only the shout of the ship's captain saved him.

Indians

As a naturalist, Maximilian duly noted the trees and plants of North America—'the beautiful yellow-headed *Icterus xanthocephalus*', the black oaks and cedars—as well as the bison, antelope, deer and fowl he increasingly saw as the *Yellow Stone* left civilisation behind. (However, even on the wild Missouri, game was getting a little scarce. Maximilian noted the story of an army hunter who had led 'three companies of riflemen' on a hunt and killed 1800 animals 'and wounded, perhaps, as many more of these animals, which they were unable to take'.)

But, as in Brazil, the prince's real interest was in the Indians. He began to pass the great Indian nation of the Omaha in early May. Near the mouth of the La Platte River, the boat stopped and the Omaha came onboard: 'The dress was of red or blue cloth, with a white border, and cut in Indian fashion. Their faces were broad and coarse, their heads large and round, their breasts pendent, their teeth beautiful and white, their hands and feet delicate.' As Bodmer sketched frantically, Maximilian watched a father place paint on the face of his son. 'He took vermillion in the palm of his hand, spat upon it, and then rubbed it on the boy's face. The head of this boy was shaved quite smooth, excepting a tuft of hair in front and another at the back.'

That night, the Omaha performed a ceremonial dance for him. One man, stripped to the waist, swung a war club 'ornamented at the handle with the skin of a polecat'. As he danced in a frenzy, other Omaha shouted 'Hi! Hi! Hi!' or 'Hey! Hey! Hey!' Maximilian wrote in awe: 'The bright light of the moon illumined the extensive and silent wilderness.' Ravens called, echoing the shouts of the Indians. It was just what Maximilian had come to America to see.

There would be more. By the end of May, the *Yellow Stone* entered the territory of the much-feared Sioux, where Bodmer sketched huge herds of buffalo. Along the riverbank Sioux dead lay buried on high wooden scaffolds, wrapped in buffalo robes, their bows and arrows beside them. Not all the dead were afforded scaffolds, however— some of the lesser Indians were merely placed in trees. As Maximilian wrote: 'We saw, in the neighborhood of this place, an oak, in which there were three bodies wrapped in skins. At the foot of the tree was a small arbor or shed, made of branches of poplar, which the relations had built for the purpose of coming to weep over the dead.'

The country was turning more beautiful and more savage. A forest fire leaped 30 metres (100 feet) in the air, blown by hot winds. Tempestuous thunderstorms pelted the river and the travellers, forcing the boat to take shelter close to the banks.

The *Yellow Stone* reached the large trading post at Fort Pierre in early June, and Maximilian, Bodmer and Dreidoppel transferred to a new steamboat, the *Assiniboine*, and headed deeper into country that was getting ever more wild. Previously, much of the game had been scared off by the smoke and clamour of the steamboat, but here wolves, swans, beavers and deer came down to the bank, filled with curiosity. Maximilian was unhappy to see men aboard the steamship 'blazing away at anything that moved on shore for the sheer sport of it'.

Going farther and farther west, the *Assiniboine* stopped at isolated trading posts, where Bodmer painted the Indians and wildlife he saw. By mid-June they had reached Fort Clark (near current-day Bismarck, North Dakota) where hundreds of Mandan and Hidatsa Indians came down to greet the steamer, carrying weapons 'such as muskets, bows, war clubs, and battle axes' and attired in buffalo skins. They were friendly—the Mandan tribe had some three decades before saved the lives of the members of the Lewis and Clark expedition by allowing the whites to winter with them—and there was much vigorous hand-shaking, which seemed to tire Maximilian out. Other Indians he met were the Crows, 'remarkably tall and handsome men' on horseback, who, he said, 'looked down' on whites. However, Crow women did assist Maximilian, Bodmer and Dreidoppel when, to their horror, they were surrounded by 'from 500 to 600' half-feral dogs who lived in the Indian encampments—the Crow squaws showed them how to throw stones to keep the animals away.

'Not a trace of the head'

On 24 June the *Assiniboine* arrived at Fort Union, at the junction of the Missouri and Yellowstone rivers, about 2900 kilometres (1800 miles) up the river from St Louis. This was as far as steamboats ventured. Maximilian and his men waited for two weeks for the arrival of the *Flora*, a big, heavy keelboat—essentially a large barge with sails and a deep keel to keep the boat stable when working against a river's current. When the *Flora* did arrive, they headed still farther up the Missouri, towards Fort McKenzie, about 800 kilometres (500 miles) away, along with about fifty *engagés*.

This arduous journey took five weeks. Slow as the steamboats' progress had been, the *Flora*'s was worse; when the wind failed, it often had to be hauled by long ropes pulled by men who walked along the banks of the river. The rain poured on the open decks of the keelboat and mosquitoes were bad. To pass time, the *engagés* hunted and soon every available surface was covered with decomposing animal corpses. When Maximilian laid out his specimens to dry, the trappers threw them overboard, claiming they took up too much room.

Finally, on 4 August, the *Flora* reached Fort McKenzie, in what is today western Montana, the most remote outpost of the American Fur Company. The outpost had existed only a year, having been built in the hopes of stimulating fur trade with the Blackfoot Indians, a tribe that had opposed white expansion in the area. When Maximilian stepped off the boat, hundreds of Blackfoot warriors were there to greet him, firing their guns in the air. When they saw how well Karl Bodmer drew, they flocked around him, begging for their portraits. Once again, as he did in Boston, Maximilian reproved the Americans in his journal for their treatment of the Indians. 'We could see drunken persons everywhere,' he wrote. 'These Indians who had drunk too much whiskey became exceedingly affectionate, shaking hands without end and even embracing and kissing us heartily. They traded their furs for whiskey and clamored for it incessantly.' Maximilian wrote that while Blackfoot men punished adultery harshly (by cutting off the offending woman's nose) they were willing to sell their wives to whites for whiskey. Maximilian was disgusted—but still traded whiskey to the Blackfoot in return for two tame bears, which he wanted to add to his collection.

Even with their drunkenness, the Blackfoot were fearsome. The prince remarked that they wore Spanish crosses and carried compasses and rifles 'marked with the names of their owners', which showed that they had taken them in war. He had heard that in 1832 alone, the Blackfoot had killed fifty-eight whites, most of them trappers or hunters wandering alone through the mountains. The Blackfoot gathered around Fort McKenzie did not, for the most part, try to kill the whites, since they traded with them, but the whites were not the Blackfoot's only enemy. Other Indian tribes that hated them were the Cree and the Assiniboine, who lived nearby.

On the morning of 28 August the prince awoke to the sound of gunfire. Dressing hurriedly, he, Dreidoppel and Bodmer loaded their guns and raced outside, gathering in the central courtyard of the fort. It turned out that a war party of six hundred Assiniboine and Cree were in the process of attacking the Blackfoot who were camped in about twenty teepees outside the fort. Most of the Blackfoot warriors were sleeping off hangovers, and so were caught by surprise when the enemy warriors slashed open their tents with knives and fired muskets and arrows inside. Looking over the wall, Maximilian saw four women and several children lying dead near the fort's walls, while the Blackfoot warriors clamoured to be let inside the gates.

The whites opened the fort for the Blackfoot, and both Indians and soldiers opened fire on the Cree and Assiniboine who raced back and forth on horseback, shooting back at the fort. In the meantime, Blackfoot reinforcements arrived from another camp a few kilometres distant and finally drove off the attackers. As Bodmer hurriedly sketched the melee, becoming the first white artist to paint an Indian battle from life, Maximilian was privy to horrifying, but fascinating, scenes. A drunken Blackfoot warrior named White Buffalo, wounded in the back of the head, sang incessantly in order to ward off the evil spirits that might kill him. An old man, wounded grievously in the knee by a musket ball, allowed a woman to cut it out with a penknife 'during which operation he did not betray the least symptom of pain'.

As the fighting moved away from the fort, Maximilian, ever the scientist, sought to examine the body of an Assiniboine who had been killed during the fray, wanting to obtain the skull as a specimen. But by the time he got there, the Blackfoot were venting their rage on the man's body. 'The men fired their guns at it; the women and children beat it with clubs, and pelted it with stones, the fury of the latter was particularly directed against the privy parts ... Before I could attain my wish, not a trace of the head was to be found.' Not to be deterred, he later took his skull specimen from an Indian burial platform.

Return to Europe

Maximilian had hoped to continue on to the Rocky Mountains, but the killings he witnessed at Fort McKenzie—the Indian battle he saw was only the most egregious example of the almost daily violence that occurred between Indians and Indians, and Indians and whites—dissuaded him. He decided to decamp back to Fort Clark and in September left Fort McKenzie for good on the 1500-kilometre (1000-mile) voyage, now travelling downstream in an oared boat that held four *engagés* as well as himself, Bodmer and Dreidoppel. The boat was stacked high with crates containing the prince's specimens and included a cage that held the two small bears he had bought.

It was a difficult voyage. They were plagued by rainy weather and each night had to sleep on shore, where they were vulnerable to Indian attack. But they finally reached Fort Clark just as winter was setting in. They spent several months there, living in the ferocious cold but studying the Mandan and Hidatsa Indians, with Bodmer making numerous brilliant studies of them. Food was scarce—at one point, the inhabitants of the fort were forced to eat a rotting elk that had floated down the river—and the prince came down with scurvy, which kept him in bed for weeks. He might have died, except for the fact that some Mandan children gave him wild onion bulbs, which acted as an anti-scorbutic.

Finally, as the ice broke up in the spring, the prince and his companions left the fort and headed back east, with most of his specimens to be sent later on another boat. Unfortunately, these were destroyed when the boat caught fire, but Maximilian still made it back to Europe in July of 1834, two years after embarking for America, with a sizeable collection of specimens that included, astonishingly, the two live bears, which were perhaps equally astonished to find themselves living in a zoo in the grounds of a castle on the Rhine.

Maximilian's two-volume *Travels in the Interior of North America* appeared in 1839, with Karl Bodmer's aquatint illustrations, but it was an expensive volume that did not reach a wide audience. And, ultimately, it was not the prince's observations about natural history that make his work so important today, but his study of Native American peoples, particularly the Mandan, who were to almost disappear off the face of the earth only three short years after Maximilian had left them, during the dreadful smallpox epidemic of 1837–39. Without the prince's journal and Karl Bodmer's paintings, our understanding of them would have been fragmentary at best.

The great smallpox epidemic of 1837

Far more than bullets or the encroachment of white settlers or the killing of game by white hunters, disease was the primary cause of the demise of American Indians. And chief among the diseases brought to America by Europeans was smallpox.

There is no way of knowing exactly, but the native population of North America has been estimated at perhaps twelve million in the sixteenth century.

By the mid-nineteenth century, it was probably something like half a million. Smallpox was in the main responsible for this. The first major outbreak in North America (the disease had already taken a devastating toll in Mexico and Central and South America in the early sixteenth century) took place in what is now New England, on the northeastern Atlantic coast, from 1616 to 1619. The Algonquin nations who lived in the area were, in those three years, reduced from thirty thousand to three hundred. When the English Pilgrim settlers arrived a year after the epidemic, they were essentially landing in a graveyard—in fact, it was only because of the decimation of these tribes that the English even survived.

After this, raging fires of smallpox rose and fell throughout the southwestern parts of America, killing hundreds of thousands. So far, however, the more remote western parts of the country had been immune, although there were outbreaks among the Indian tribes along the Pacific Coast in 1780 and 1802. However, in 1832, just before Maximilian of Wied made his journey, the first steamboat sailed up the Missouri to Fort Union, opening up the area to more and more travellers. Disease outbreaks were almost inevitable. In June of 1837, a steamboat named the St Peter, *carrying men infected with smallpox, made its way up the Missouri. Officials of the American Fur Company tried to keep the Mandan Indians away from the boat but were unable to—American Indians had no understanding of how infectious disease was spread.*

Smallpox rapidly found its way into two Mandan villages; only thirty-one Indians survived out of a total of sixteen hundred. The Blackfoot and Assiniboine suffered the same fate—the disease had a ninety-eight per cent fatality rate among people who had developed no immunity to it. Smallpox was also spread by warring Indians who killed those infected with smallpox or raided the empty villages of the dead, taking home infected blankets and other belongings. A trader described the western landscape: 'In whatever direction we go, we see nothing but melancholy wrecks of human life.'

It is difficult to know how many of the Indians Maximilian of Wied had visited died—estimates range from fifteen thousand to sixty thousand—but entire native cultures disappeared in a three-year span from 1837 to 1840.

CONSTANTINE SAMUEL RAFINESQUE
'It is usually safe to reject his conclusions'

Turkish-born naturalist Constantine Samuel Rafinesque was one of the most annoying human beings most people introduced to him had ever met. In between irritating people, however, he got quite a lot done. This 'relentless scrivener', as one biographer has called him, published more than a thousand books and articles on all manner of topics. As he himself wrote—Rafinesque was always his own best publicist—he was 'a Botanist, Naturalist, Geologist, Geographer, Historian, Poet, Philosopher, Philologist, Economist, Philanthropist ... Traveler, Merchant, Manufacturer, Collector, Improver, Professor, Teacher, Surveyor, Draftsman, Architect, Engineer, Pulmist [Pulmonologist], Author, Editor, Bookseller, Librarian [and] Secretary'.

And, at least if you believe Constantine Rafinesque, he was brilliant in all these areas. 'I never fail to succeed,' he wrote, 'if depending on me alone, unless impeded and prevented by ... the hostility of the foes of mankind'. These 'foes of mankind' included those other scholars and naturalists whose 'endless discrimination' (once again, according to Rafinesque) kept him from receiving the fame that should have been his just reward. These enemies included the Philadelphia journalist who wrote: 'It is quite fitting that the name "Rafinesque" rhymes with "picturesque" and "grotesque," because so the little man is.' Or the Baltimore historian Brantz Mayer, who simply wrote: 'It is usually safe to reject [Rafinesque's] conclusions.' Or the famous ornithologist John James Audubon, who described Rafinesque as 'an odd-looking fellow' and played a famous practical joke on him.

And yet this 'self-centered, stubborn, quarrelsome' man (as another contemporary depicted him) has also been described 'as one of the most remarkable [naturalists] to appear in the annals of American science'. His thoughts on the impermanence of species predated Darwin's theory of evolution by a generation. He discovered and classified hundreds of American plants and animals. And he also understood the significance of fossils in dating geological strata, highly important when it comes to figuring out the age of the earth.

'To enjoy life'

Constantine Samuel Rafinesque was born in 1783 in a Constantinople suburb to a French trader father and a German mother. When he was an infant, his parents moved back to his father's home, Marseilles. The ocean journey from Turkey to France supposedly protected Rafinesque forever from seasickness, as this excerpt from his autobiography, *A Life of Travels in North America and South Europe*, attests:

> *This first and early voyage of mine, made me insensible or not liable ever after to the distressing seasickness. By my observations ever since, it appears that whoever travels by sea in the cradle or very early, is never liable afterwards to this singular disorder. It seems also that whoever can ride backwards in a coach without difficulty, is not liable to it; but whoever cannot, will suffer sadly from it. Thus any one can*

tell beforehand whether he will be liable or not, and take the needful precautions. Having never read anywhere these two observations of mine, I venture to notice them here, that they may be confirmed by extensive experience.

It is worth including this extract as an early example of what irritated Rafinesque's critics so much—his drawing confident, sweeping conclusions based on little or no evidence.

Rafinesque lived in Marseilles until he was nine, although his father was absent most of the time, travelling through Europe and North America on business. In the early 1790s, with the French Revolution targeting wealthy people like the Rafinesques, his mother moved Constantine and his brother and sister to Leghorn, Italy, where her husband's parents lived. There, Rafinesque, a small, dark-haired, frenetic boy, began to wander the woods around his home, pursuing his love of 'flowers and fruits'. He was just beginning, as he would write, 'to enjoy life' when news came in 1793 that his father had died during a yellow fever epidemic while visiting Philadelphia. In 1797 his widowed mother moved the family back to Marseilles, where Rafinesque—'greedy for reading any book I could get'—was privately tutored and claimed that he learned fifty languages. But then his wealthy grandparents died, Rafinesque's inheritance turned out to be nearly nothing—the first, but not the last, financial disappointment of his life—and the young man needed to find a profession.

In 1802, when he was eighteen, Rafinesque set out for America with his younger brother, Anthony, buoyed by a sense of adventure and the promise of a job at a Philadelphia 'counting-house', or accountant's firm, a job secured through connections of his late father.

Draba Americana

Arriving in Philadelphia in mid-April, Rafinesque tried for a short time to work diligently at the counting house, but the lure of nature and the vast American wilds called to him. Within a few months,

Anthony took over his job and Constantine left to make 'excursions into the virgin Botanical world of Pennsylvania and surrounding states'. It is typical of Rafinesque's conceit to call these areas 'virgin'; they had been explored by naturalists like John and William Bartram (see page 66), among others. In fact, one of the first 'new' plants Rafinesque saw and grandly classified as *Draba Americana* was, in fact, *Draba verna*, a common weed long known in America and Europe.

But Rafinesque, with admirable energy and a touch of the mania that would characterise all his work, roamed for hundreds of kilometres through Pennsylvania and New Jersey, meeting his first Indians (the Delaware tribe)—whom he says he instantly identified as being of 'Tartar or Siberian origin'—and building a collection of hundreds of plants, as well as birds and reptiles, particularly snakes. Rafinesque's observations of the latter are astute and imaginative, showing that when he was not trying to compete or outdo others, he could be a brilliant naturalist. In fact, many scholars consider him one of America's first herpetologists.

By 1804, botanising with his brother (who was now out of a job), Rafinesque became acquainted, as he wrote, 'with all the Botanists, Naturalists and Amateurs of that period'. Some of them he charmed with his energy and enthusiasm for their work, enthusiasm that, when genuine, was often embarrassingly excessive. Through certain prominent naturalists Rafinesque was able to meet American Secretary of State (and future president) James Madison and, through Madison, President Thomas Jefferson, himself a man with an extensive interest in the natural world. An exchange of letters followed. Rafinesque hoped to join the famed Lewis and Clark expedition that Jefferson was then forming in order to explore the western territories newly acquired in the Louisiana Purchase, but seeing that even such a famous naturalist as Alexander Wilson (a noted ornithologist) had been turned down for the post, he did not think he had much of a chance and so did not bother to apply. Instead, discouraged at his lack of advancement, he decided to sail back to Leghorn in late December 1805. It was bad timing. A few days after Rafinesque left, Jefferson wrote to tell him that he might be able to offer him a position on another westward-bound expedition—but by that time Rafinesque found himself at sea.

Of squill and women

Constantine Rafinesque thought that his fortunes might improve in Europe and, temporarily, they did. Deciding to make his home in Palermo, Sicily, Rafinesque sold his collection of plants, animals and reptiles, and rocks (ten thousand specimens, or so he claimed) for a fair sum. Because of the Sicilians' extreme dislike of Napoleon and the damage caused by the Napoleonic Wars, Rafinesque decided that French was not a good thing to be. Therefore, he changed his name to Constantine Samuel Rafinesque Schmaltz, Schmaltz being his German mother's maiden name. To further distance himself, and to make a living, he got a job as secretary to the US consul in Palermo, grandly signing his own correspondence 'Chancellor of the American Consulate'. This position left him time for botanising, and for the first time he also used his knowledge of plants to make money—he cultivated squill, which are the dried bulbs of the sea onion, and sold them in bulk as an expectorant and diuretic. Buying for a dollar per hundred pounds (45 kilograms) and selling for up to thirty dollars for the same quantity, Rafinesque made a killing.

In 1809, during this brief period of happiness and good fortune, Rafinesque met and married a Sicilian woman, Josephine Vaccaro, although the marriage was not sanctioned in Italy, possibly because Rafinesque was not a practising Roman Catholic. They had a daughter, Emily, in 1811 and a son, Charles Linnaeus—named after Rafinesque's great hero, Carl Linnaeus—who died as an infant in 1813. But the marriage did not go well, to say the least. In all his writings, Rafinesque did not speak much of family and personal matters—possibly because he found these subjects too intimate and more likely because of his narcissistic tendencies—but he did later call Josephine 'false' and 'stupid'. This may have been because when, as we shall see,

Rafinesque was shipwrecked returning alone to America, Josephine, without even waiting to hear whether he was alive, quickly married a vaudeville comedian and ran off with him, taking Emily with her. Rafinesque never saw wife or daughter again.

By 1815 Rafinesque was sick of Sicily, which he began calling the 'land of perfidious women'. (One can perhaps assume that Josephine had had an eye for the comedian even before Rafinesque left.) Things had gone badly. The rest of Sicily had caught up to him in the production of squill, causing the bottom to drop out of the market. Rafinesque's self-published works of botany had not brought him the reputation he craved and, along with it, an appointment as a professor at the University of Sicily. Deciding to leave the country to 'vegetate in its willful ignorance', he set sail for America aboard the ship the *Union of Malta* in July. With him he carried everything he owned. As he wrote: 'Besides 50 boxes containing my herbal, cabinet, collections and part of my library. I took all my manuscripts with me, including 2000 maps and drawings [and] 300 copperplates. My collection of shells was so large as to include 600,000 specimens large and small.'

'I hope we may be of use to each other'

The trip to America, with stops at Gibraltar and in the Azores, took over a hundred days, much of it in stormy weather, by which time Rafinesque was heartily tired of being onboard, despite his supposed immunity to seasickness. When the *Union of Malta* arrived off Long Island in early November, it was met by fierce westerly winds and a thick fog and ran aground between Fisher's Island and Long Island. Through great good fortune, passengers had time to escape in lifeboats, rowing towards the beam of the lighthouse in New London, Connecticut, but everything Rafinesque had was lost.

As if this were not a dramatic enough experience, Rafinesque had to embellish it. Quite preposterously, he later told a friend that as he swam away from the shipwreck to a waiting lifeboat, he was able to observe and identify several new species of plants and fish—such an extraordinary claim, given the fact that the wreck happened at night and in the fog, that Rafinesque's recent biographer Leonard Warren postulates that 'at times of extreme anxiety, [Rafinesque] seems to have been delusional'.

In a self-described state of 'utter despair', having lost all his possessions (and soon to lose his wife and daughter) Rafinesque walked from New London to New York City, where he prevailed upon an influential physician friend from his previous days in America, Dr Simon Mitchell, to help him find work. He took a position as a tutor to a wealthy family but lost the job when the family moved. Once again homeless and out of work, Rafinesque was taken in by another old friend, a Philadelphia Quaker named Zaccheus Collins, who apparently gave him a stipend while he went out botanising and hunting for fossils and reptiles. Sometimes he went with Mitchell and other plant hunters, more often he travelled alone, and on foot. He despised horses for botanising—'I never liked horses and dismounting for every flower; horses do not suit botanists'—but he covered a great deal of ground, in 1816 alone wandering almost all the way up the Hudson River, mapping it as well as collecting.

Within a few years Rafinesque had become a member of the Academy of Natural Scientists and had begun publishing numerous books and articles. One 1816 article, entitled *Circular Address on Botany and Zoology*, proposed that collectors everywhere exchange plant specimens with each other, creating a worldwide fraternity of science: 'Whatever be your situation in life and whatever is your abode, I hope we may be of use to each other ... Ask me in return to bestow anything in my power to bestow, plants, animals, books, my works &c.'

This plan, while idealistic, was ignored and Rafinesque quickly moved on to what he considered to be a major work: his *Annals of Nature and Somiology of North America*, to be serially published over ten years, containing descriptions of all the plants and animals in North America. Obviously a tall order, and most scientists resented and ridiculed him for it, in part because Rafinesque often announced such grand plans but never followed through on them (as indeed would be the case with *Annals of Nature*). But the main issue was his attitude. Many of his plant and animal 'discoveries' had been previously discovered; Rafinesque merely used a different system of nomenclature (building on the strict Linnaean system of sexual identification used by most American naturalists) to describe their genus and species. Amos Eaton, an American botanist who was actually fond of Rafinesque, wrote to a friend in some exasperation: 'Why can not he [Rafinesque] give up that foolish European foolery, that leads him to treat Americans like half-taught schoolboys?'

'Never mind, Mr. Audubon'

In the spring of 1818, Rafinesque began a 3000-kilometre (2000-mile) journey over the Allegheny Mountains, part botanising trip, part surveying, for he was also creating a chart of a portion of the Ohio River for a Pittsburgh bookseller. That summer the river brought Rafinesque unannounced to the home of John James Audubon, who had not yet become famous as an ornithologist and bird painter and was making a living running a store and flour mill in Henderson, Kentucky. Yet his work with birds was certainly known in naturalist circles, and one fellow scientist had given Rafinesque a letter of introduction. Rafinesque's visit to Audubon provides both high comedy and an unparalleled glimpse of the French naturalist in action.

Audubon, with little idea that he would soon meet the man, spotted Rafinesque as soon as the latter walked off the boat. He thought to himself 'What an odd-looking fellow!' a few moments before Rafinesque approached him, identified himself and handed him the letter of introduction. Audubon kindly took Rafinesque to his home and offered to send for his luggage. 'He told me he had none but what he brought on his back', Audubon wrote in his journal. 'The naturalist pulled off his shoes, and while engaged in drawing his stocking, not up, but down, in order to cover the holes about the heels, told me in the gayest mood imaginable that he had walked a great distance.'

Audubon fed his 'ravenous guest' dinner and had an 'agreeable' conversation with him, all the time shaking his head inwardly at Rafinesque's 'singular appearance':

A long loose coat of yellow nankeen, much the worse for the many rubs it had got in its time, and stained all over from the juice of plants, hung loosely about him like a sack. A waistcoat of the same, with enormous pockets, and buttoned up to his chin, reached below a pair of pantaloons, the lower parts of which were buttoned down to his ankles. His beard was as long as I have known my own to be during my peregrinations, and his lank black hair hung loosely over his shoulders.

Despite the way Rafinesque looked, Audubon initially found him good company, an intelligent man who made 'delightful' conversation. But he was decidedly odd. After dinner, he examined some of Audubon's sketches and found 'a plant quite new to him'. Audubon told him the plant was common nearby and he would show it to him tomorrow, but nothing would do but that Rafinesque should see it that evening. Audubon relented and took him to the riverbank, where he pointed out the plant. At this point Rafinesque 'plucked the plants one after another, danced, hugged me in his arms, and exultingly told me that he had got not merely a new species, but a new genus'.

Later that night, after his guest had retired, Audubon heard 'a great uproar' coming from Rafinesque's room. Thinking something terrible had happened, he raced to his guest's door and pushed it open. 'I saw my guest running around the room naked,' wrote Audubon, 'holding the handle of my favorite violin, the body of which he had battered to pieces against the walls in attempting to kill the bats, which had entered by the open window, probably attracted by the insects flying around his candle'.

Rafinesque was not afraid of the bats; he merely wanted specimens of what he considered to be another 'new species'. Audubon obligingly killed a few for him, but then complained about the fate of his violin.

To which the French naturalist replied maddeningly: 'Never mind, Mr. Audubon ... I have my bats, and that's enough.'

Rafinesque was to stay for three more weeks and then simply disappeared one morning without saying good-bye. But not before Audubon, who never got over the loss of his violin that first evening, decided to get his revenge by making up fantastical American natural creatures (even drawing them) and presenting them to the wide-eyed Rafinesque as if they were real.

'Never-failing lack of tact'

Having finally finished his journey in Philadelphia at the end of 1818, Rafinesque proudly proclaimed that he had gathered 'Abt. 25 new species of Bats, Rats and other quadrapeds, abt. 20 N. Sp of Birds, Abt. 15 N. Sp of snakes, turtles, lizards, and other reptiles, 64 N. Sp of fishes of the Ohio, more than 80 N. Sp of shells, besides some new worms and many fossils'. This was in addition to six hundred plant specimens, of which sixty were new species.

Fellow naturalists looked askance at the plethora of new species Rafinesque claimed that he had collected, but this did not stop him from parlaying his journey into a travel memoir and a new job: Professor of Natural Science at Transylvania College in Lexington, Kentucky, an early college known as 'the pioneer college of the western wilderness'. Rafinesque, in fact, became the first professor of science west of the Allegheny Mountains. And he did well—his eccentricities, which irritated his fellow naturalists, entertained students, and he had the habit, unusual for the day, of lecturing from live specimens rather than dusty books. For the next seven years, Rafinesque taught at Transylvania, loved by students but clashing constantly with the administrators, whom he considered to be 'foes of science' but who were offended, in the main, by what one contemporary called Rafinesque's 'never-failing lack of tact.'

The job in any event paid little more than room and board, and he was forced to scramble for money. He put ads in the local paper, offering to teach French to Kentucky ladies. He also advertised to show people how to 'compel' mussels (which abounded in local waters) to form pearls. But even as he fought to stay afloat financially, Rafinesque's energy kept him out on botanising trips. With private subscriptions, he also created a public botanical garden that would eventually include up to forty thousand plant and animal specimens. He also invested a good deal of time studying American Indians. It was probably at this time that he began writing the *Walam Olum*, the 'translation' of the creation myth of the Lenni Lenape Indians that would confuse scientists for a hundred and fifty years.

Eventually, in 1826, his relationship with the president of Transylvania College, Horace Holley, became so strained that he was forced to leave, not without, however, putting curses on both the school and Holley. As Rafinesque wrote with vengeful satisfaction: 'Leaving the College with curses on it and Holley; who were both reached by them soon after, since he died next year at sea of the Yellow fever, caught at New Orleans, having been driven from Lexington by public opinion: and the College has been burnt in 1828 with all its contents.' (Holley did indeed die within a year of Rafinesque's departure, but the school, although it suffered a serious fire, did not burn to the ground.)

Audubon and Rafinesque: 'fake fish foolery'

John James Audubon, America's great patron saint of ornithologists, was at first fascinated and somewhat charmed by the odd character of Constantine Samuel Rafinesque, but he soon grew tired of his importunate houseguest and his pretensions, as many had before him. He decided to play a trick on him.

Rafinesque delighted in examining Audubon's drawings of various bird, plant and fish specimens, and so Audubon began to insert sketches of fish he had entirely made up. He drew a large, sturgeon-like creature with no ventral fins, a fish with an improbably huge head, and as many as eight other fish, all of them equally preposterous. One of them was called 'The Devil-Jack Diamond Fish' and was, according to Audubon, 4 to 10 feet

(1.2–3.0 metres) long, weighed four hundred pounds (180 kilograms), and was bulletproof because its body was covered with 'stone scales'. Rafinesque was amazed and delighted. He told Audubon that these fish were completely new species, but his host demurred modestly, claiming that the creatures were quite common in the Ohio River.

After Rafinesque left, Audubon chuckled at the joke he had played on the eccentric naturalist, but the joke turned out to be on him. When Rafinesque published his book Ichthologia Ohiensis, *about fish of the Ohio River, it contained every fake species Audubon had drawn for him, including exact descriptions, names and drawings. And Audubon was given credit for their discovery. In the case of the sturgeon-like fish, Rafinesque wrote: 'This genus rests altogether upon the authority of Mr. Audubon, who presented me with a drawing of the only species belonging to it.'*

There is no doubt that Rafinesque believed Audubon and was simply trying to give credit where credit was due, but the joke hurt both men's reputations, since Rafinesque's otherwise serious book was called into question. And, a number of years later, when Audubon was trying to get his own ornithological work into print, critics claimed that if he could make up outlandish fish, he could make up birds too. Audubon admitted in a letter to a friend that his 'fake fish foolery' had cost him a good deal.

Envious hearts

Forty-three years old, turning fat and bald, Rafinesque went back east to Philadelphia, where he fell on hard times. He taught geography in a local high school but found this too 'arduous'. He attempted to take out patents on various inventions, among them a steam plough, an aquatic railroad and what he called 'the Divital Invention', which was a method for issuing bank stock in easily traded units. None of these ideas went anywhere; most were, in fact, stolen—according to Rafinesque—by unscrupulous people. 'I was compelled to foil this kind of swindling or knavery,' Rafinesque wrote, 'by not taking any more patents; but using secretly my Inventions. Some envious hearts may have blamed me for it: they are probably those who would have been the first to steal them if published.'

It seems evident that Rafinesque, always given to suspicion and paranoia, was falling more and more under their influence as his life became more difficult. He turned to patent medicine, offering consumption cures via 'a chemical manufacture of vegetable remedies'. Every year his account books show him falling deeper and deeper into debt. Yet he continued to think and work and, when possible, take botanising trips. In 1833 he wrote in one article that 'there is a tendency to deviations and mutations through plants and animals by gradual steps at remote irregular periods ...' He was describing evolution, the first scientist in America to do so, but no one paid any attention.

Towards the end of his life, Rafinesque wrote his *cri de coeur*, a book-length poem entitled, fittingly from his point of view, *The World, or Instability*. One verse reads: 'Still in hope we trust, and love we feel / Since if deprived of both we should become / Unfeeling wretches.' Perhaps it was hope and trust that kept Rafinesque going: it is difficult to say. He spent his last few years in a rented house in Philadelphia, before dying of stomach cancer in 1840, at fifty-six years of age.

Despite the fact that his death was nowhere near as squalid and dramatic as has been reported, he died deeply in debt, with most of his belongings consigned to the rubbish tip. Gradually over the years Rafinesque's reputation has been rehabilitated, although he remains, as one historian has put it, 'the most enigmatic and controversial figure in American Natural history'. He is credited with having discovered thirty-five genera and thirty species of fish, six genera and six species of animals, and some seventy genera and seventy species of plants. Near the end of his life, complaining as usual, Rafinesque wondered: 'Why should I not [have found] protectors or enlightened patrons, as were found by Audubon and others?' The answer lay within himself, in his own peculiar form of madness. However, in the same breath he stated that his main mission had been to 'enlarge the limits of knowledge' and in this, at least, he succeeded.

Stranger's Ground

The morbid legend of Rafinesque's death in Philadelphia in 1840 has him passing away painfully from stomach cancer, alone in a squalid garret. And, to make matters worse, Rafinesque's body was then held for ransom by his landlord in an attempt to collect unpaid back rent. If the rent were not paid, the man told Rafinesque's few friends, he would sell the naturalist's body to a medical school for dissection. The legend goes on to state that, to avoid that sordid end, two friends of Rafinesque's broke into his attic room, lowered the body by ropes to the street and made off with it for a private burial.

This fits nicely into Constantine Samuel Rafinesque's legend, but it is almost certainly not true. While it is true that Rafinesque died a pauper, he was attended, according to his biographer Leonard Warren, by two very able physicians who took care of him as an act of charity because of his standing as a natural historian. Another doctor, a personal friend named James Mease, was also present. There is little evidence that Rafinesque's landlord held his corpse for ransom.

Still, there is pathos and irony aplenty to be found in the naturalist's death. Mease, his executor, had Rafinesque buried in Ronaldson's cemetery, which was also known as Stranger's Ground, because it was often used to bury those not native to the city and without religious affiliations. Much later, in 1924, when Rafinesque's reputation was undergoing rehabilitation, his bones were dug up and transported to the grounds of Transylvania University in Lexington, Kentucky, where he had taught. They reside there to this day.

Unfortunately, these are probably not his bones. Since it was the practice in Stranger's Ground to stack numerous bodies in a single grave, scholars have presented evidence that the bones at Transylvania belong to a woman named Mary Passmore, who died in 1848 and was buried above Rafinesque. It is most likely that the mortal remains of Constantine Samuel Rafinesque now reside beneath the asphalt of a Philadelphia playground that occupies the space where Stranger's Ground once stood.

JOHN RICHARDSON
'an inhabitant of these wilds'

Almost all the naturalists we encounter in these pages meet with great difficulties on their journeys in search of the hidden treasures of the world—they fight their way through jungles and over steppes, crawl, quite literally, into the tiger's den—but none suffer the tribulations of John Richardson, the Scottish surgeon-naturalist who accompanied Arctic explorer John Franklin on his first journey into the Canadian wilderness in search of the Northwest Passage. Stranded in the vast and desolate regions of the far north as winter set in, Richardson was desperately trying to keep a fellow traveller alive when his campsite was visited by a French-Iroquois voyageur named Michel Teroahaute. Teroahaute claimed that he had come to save Richardson and his companions, but it soon became apparent that the man, who wandered in and out of camp raving, was dangerously mad. Within days, Richardson was faced with a situation that few of our naturalists have had to encounter: to kill or be killed.

'On the whole rather plain'
If there was one man up to confronting such a dilemma, it was John Richardson. Richardson was born in Dumfries, Scotland, in 1787,

the oldest of twelve children of a brewer and his wife. He showed an early aptitude for academics, learning to read at the age of four, and when he was thirteen he was apprenticed to an uncle who was an Edinburgh surgeon. Richardson graduated from the Royal College of Surgeons at the age of nineteen, in early 1807, and then signed on as an assistant surgeon with the Royal Navy, where he served against the French in the Napoleonic Wars and against the Americans in the War of 1812, by which time he had become a full surgeon.

Being a naval surgeon was not what anyone would call easy duty. Richardson spent a good deal of his time in combat operating without anaesthesia on screaming men in the small, dark quarters of a wildly manoeuvring ship. By all accounts, he saved many more lives than he lost. On half-pay leave from the navy after the War of 1812 reached its conclusion, Richardson went back to Edinburgh and resumed his medical studies, writing a thesis on yellow fever and becoming a full doctor in 1816. He married Mary Stiven in 1818 and practised medicine in Leith, the port of Edinburgh, while also studying natural history at the University of Edinburgh.

On the surface, Richardson, while obviously talented and successful, was an ordinary-enough looking fellow. John Franklin's future wife, Jane Griffin, would describe him as 'a middle-sized man ... not well-dressed—looks like a Scotsman as he is—he has broad & high cheekbones, a widish mouth, gray & brown hair, on the whole rather plain, but the countenance thoughtful, mild and pleasing'. But beneath his placid surface, Richardson was a man who sought adventure and who felt trapped in the life of a genteel Edinburgh doctor. Fortunately, a chance to escape was just about to come his way.

The Barren Lands

After the defeat of Napoleon at the Battle of Waterloo in 1815 marked the end of the Napoleonic Wars, Great Britain was left with a large and expert navy that had very little to do. The obvious solution to such underemployment was for the Royal Navy to resume its great tradition of seafaring exploration, as exemplified by the eighteenth century by Captain James Cook and others. One object of such exploration was a resumed search for the Northwest Passage through the mainly unexplored Arctic seas of North America. In 1818 separate expeditions under Captain John Ross, and Captain David Buchan and Lieutenant John Franklin surveyed Baffin Bay and unsuccessfully sought a passage far to the north.

Shortly after this, the British Admiralty decided to send John Franklin out on an overland expedition to map Canada's coastline northwest of Hudson Bay. This was a vast area often called the Barren Lands and usually marked *Terra Incognita* on maps—the area between the Coppermine River and, 800 kilometres (500 miles) to the west, the Mackenzie River, both of which emptied into the Arctic Ocean. Very few Europeans had visited this land, which was occupied in the far north by Inuit tribes and to the south by Chipewyan Indians, two Native American peoples who warred ferociously with each other.

The idea was for Franklin to map the coastline between these two rivers (or, if circumstances dictated, the coast east of the Coppermine), to study the temperatures and air quality, and to get a sense of the animal life and the inhabitants, all of which would prove highly useful for future explorers seeking the Northwest Passage. Among other experts, Franklin would need both a doctor and a naturalist and he got both in one person: John Richardson. In early 1819, Richardson was approached by the Secretary of the Admiralty, who wrote: 'If you feel disposed for such an expedition, and think that your health and qualifications are suitable for the undertaking, and you could be ready to set out from England by the first week of May, I request you to state to me whether you could undertake to collect and preserve species of minerals, plants, and animals ...'

It was not much time for Richardson to collect his supplies, put his affairs in order, explain the situation to his wife of just over a year and depart in good order, but in an eager frenzy of activity, the doctor made himself ready. As he wrote to his father just before departure: 'The country has never been visited by a naturalist and presents a rich harvest ... If I succeed in making a good collection, I have no doubt of my promotion on my return.'

'Have I this day walked with God?'

The expedition left England on 23 May 1819, aboard three ships belonging to the Hudson's Bay Company, one of the two chief fur-trading companies in Canada—the Royal Navy had purchased accommodation aboard these annual supply vessels as a cheap method of transporting the small group of six men.

Lieutenant John Franklin led the foray. Franklin, who would go on to become one of the most famous, and most tragic, Arctic explorers, was thirty-three years old and an experienced seaman and explorer, having circumnavigated Australia with Matthew Flinders in 1800 and fought in the battles of Trafalgar and New Orleans. Yet he was a surprising choice, a man so gentle that he literally hated killing even a mosquito, let alone flogging a rebellious seaman. Franklin was somewhat overweight and physically ill-conditioned, with circulation so poor that he suffered from cold hands and feet even in England. He was also fervently religious, ending each day by writing in his journal: 'Have I this day walked with God?' Aside from Richardson, the other officers on the voyage were Robert Hood, an extraordinary young artist, and George Back, a seasoned navigator and mapmaker. Rounding out the expedition were ordinary seamen John Hepburn and Samuel Wilks.

The task of these few men was enormous. Landing in Hudson Bay at York Factory (chief trading post of the Hudson's Bay Company), they were to row or portage two heavy boats hundreds of kilometres through the Barren Lands to the mouth of the Coppermine River, making stops at Cumberland House (a supply depot on the Saskatchewan River), Fort Chipewyan on Lake Athabasca, and finally the Great Slave Lake, where they would set up an outpost before striking north to the Coppermine and following it north of the Arctic

Circle into what is now known as Coronation Gulf, an arm of the Arctic Ocean. From there, Franklin could turn either east or west and map the coastline.

Naturally, everything began to go wrong as soon as possible. The British Admiralty, in its wisdom, had decreed that Franklin should stop in the Orkney Islands on the way to Canada to pick up a group of the famed boatmen of that region to pilot the boats through Canada. When Franklin got to the Orkneys, however, an unusually abundant run of herring had claimed most of the Orkneymen, and he found only four (he had wanted ten) willing to go with the expedition, and this only as far as Fort Chipewyan. Then, when the expedition finally arrived at York Factory on 30 August, they discovered that the two primary trading companies of the region, the Hudson's Bay Company and the North West Company, were engaged in warfare over the rich fur spoils of the region and were very reluctant to provide Franklin with the men and supplies he needed. A disgusted John Richardson wrote to one of his brothers: 'The contests of the rival Fur Companies have been carried on in a disgraceful and barbarous manner.'

'The prospect of return'

Despite all of these problems, the expedition left York Factory on 9 September on the first leg of its journey to Cumberland House. It was a tough baptism in the Canadian wilderness as they struggled upstream with the heavy boats—at one point they were forced to dump half of the supplies they were carrying in them, in order to make an arduous portage. The group finally reached Cumberland House on 23 October, with the winter freeze already setting in. In January, Franklin, Back and Hepburn set out on snowshoes to reach Fort Chipewyan (where they finally arrived in March of 1820). Richardson, Hood and Wilks wintered at Cumberland House. It was Richardson's first experience of the Canadian winter and despite the harshness of the climate, he found it beautiful. Walking through the forests each clear day to survey the surrounding wilderness, he found five species of birds not known to the scientific world, and Hood then captured them in paintings, although the weather was so cold his brush sometimes froze to the canvas. Still, time weighed heavily on the men. As Hood wrote in his journal: 'All the productions of the earth were buried til spring. No sound disturbed the silence of woods, but for the frequent strokes of the axe ... In such a state one might be disposed to envy the half year's slumber of the bears.'

Richardson and Hood finally joined Franklin and his party at Chipewyan in July of 1820 to prepare for the push north beyond the Arctic Circle (Wilks, whose health was apparently not good, had been discharged from the expedition and had returned to Britain in the spring). Franklin attempted to persuade the Orkneymen to go forward with the expedition into the uncharted wilderness and in his journal reports their reaction: 'they minutely scanned all our intentions, weighed every circumstance, looked narrowly into the plan of our route, and still more circumspectly to the prospect of return.' Feeling that the expedition had little chance of succeeding, they politely declined to go farther and returned to Cumberland House.

Franklin, Richardson and the rest of the British, as well as a group of voyageurs—French-Canadian trappers hired by Willard Ferdinand Wentzel, a North West Company clerk, to help guide and hunt for the British party—began their journey north. Franklin was not particularly happy with the quality of these men, who included Michel Teroahaute, but the experienced voyageurs of the region saw little profit in venturing deeply into this wild, perilous region. Also accompanying the expedition was a group of Copper River Indian hunters, under Chief Akaicho, hired to provide food for party.

Troubled journey
There was trouble right from the start as the expedition headed north. On their first night into the journey a careless campfire burned down a good part of the forest around them. It turned out Akaicho and his men were of little use as hunters, spending days out of sight and

returning with almost nothing to show for it. Because of this, when Franklin set up winter camp at Fort Enterprise on the Coppermine River, just inside the treeline before the Arctic Circle began, he was forced to send George Back to Cumberland House to get additional supplies, which Back did, making an incredible 2000-kilometre (1200-mile) round-trip journey on snowshoes.

The harsh winter of 1820–21 was spent in the draughty confines of Fort Enterprise—the grandiose name for the wooden huts Franklin and his men built to protect themselves from the elements. The voyageurs accompanying the party caused problems by grousing constantly about the cold and the lack of food, so much so that Franklin had to warn one of them, the party's interpreter Pierre St Germaine, that he could be executed for mutiny if he was returned to Britain for trial. St Germaine only laughed. 'It is immaterial to me if I lose my life,' he retorted, 'for the whole party [is going to] perish.'

To make matters worse, both Hood and Back fell in love with a young Indian girl named Green-stockings and nearly fought a duel over her affections. John Hepburn overheard them discussing this, however, and in a scene worthy of a French farce removed the charges from their pistols. (During Back's absence, Hood did get Green-stockings pregnant, and she gave birth to a child listed in a later census as 'the orphaned daughter of Lieutenant Hood'.)

In the midst of all this, John Richardson's journal, which he began keeping in August 1820, stands out as an oasis of calm. He was intent on fulfilling his job as the expedition's naturalist by closely watching the environment around him. He collected temperatures for the month of August—showing how extreme the weather could be, since the temperatures fluctuated from 0.5 degrees Celsius (33 degrees Fahrenheit) to 25.5 (78) degrees. He described the coming of the first ice, on 14 September and listed twelve different types of lichen to be found near Fort Enterprise—one, covering 'the surfaces of most of the larger stones which strew the barren grounds' was called *tripe de roche* and would become quite familiar to Richardson the following year. By 9 October, the water of the nearby Winter Lake was frozen over, and Richardson turned to describing the local reindeer, noting that 'the fat is at this season deposited to the depth of two inches [50 millimetres] or more on the rumps of the males and is beginning to get red and high flavoured, which is considered an indication of the commencement of rutting season'.

Richardson also closely describes the way the Copper River Indians hunted the reindeer—having realised that the creatures were attracted to the colour white, they would don white clothing or tie a strip of white cloth around their heads, and then 'often succeed in bringing [the reindeer] within shot by kneeling and vibrating the gun from side to side in imitation of the motions of a deer's horns when he is in the act of rubbing his head against a stone'.

By November, Richardson noted that the low temperature was down to minus 35 degrees Celsius (minus 31 degrees Fahrenheit). The reindeer had become scarce (they had headed farther south for the winter) and the Indian hunters 'remain at the house and as they require rations, they encrease greatly the consumption of provision'.

The Bad Lands

In June of 1821 the expedition left Fort Enterprise to follow the Coppermine north to the Arctic Ocean. On 14 July John Richardson led his party in climbing a promontory overlooking the Arctic Ocean, becoming only the fourth European to see it from the North American mainland (the explorers Samuel Hearne, in 1771, and Alexander Mackenzie and John Steinbruck, in 1789, were the others).

The Arctic Ocean was free of ice and Franklin now determined that he would head eastward to map the coastline as far as Repulse Bay on the northwestern shore of Hudson Bay—a journey of 1200 kilometres (750 miles) in birch-bark canoes (the heavy rowboats had long since been abandoned), a highly unrealistic goal in the short Arctic summer. Since Akaicho and his Indians refused to accompany Franklin farther because of conflicts with the Inuits, Franklin sent them back to Fort Enterprise, instructing them to kill game during the summer to stock the fort for when the expedition returned. Then he set off with Richardson and the other British, as well as eleven voyageurs.

Astonishingly, Franklin actually charted 900 kilometres (555 miles) of coastline by 18 August, when he decided to turn the expedition around. Richardson was in his element, making minute descriptions of fish in the northern waters and writing some of the very first descriptions of the yellow-bellied loon, the hawk owl, the green-winged teal drake and the red-necked grebe.

'Though deprived of bread or vegetables'

John Richardson was a superb naturalist, but he especially excelled at the study of fish, or ichthyology. C. Stuart Houston, the editor of Richardson's journals, says that Richardson identified 'forty-three still-accepted genera of fish and well over two hundred new species of fish'.

However, Richardson's interest in the finny creatures was no abstract one. He loved the taste of fish and devoted pages in his journals to describing them. 'The burbot,' he wrote, 'is found in every river and lake of this country, and is so little esteemed as food as to be eaten only in cases of necessity'. However, its roe could be baked into biscuits that Richardson found 'very good'. Richardson went on to write that the pickerel, also quite common in northern Canada, was 'a well-flavoured, delicate fish, though being too poor to please the palates of those who have been accustomed to feast upon the White fish or ... Sturgeon, it is often abandoned to the dogs, with whom, for the same reason, it is no favourite'.

Richardson saved his greatest praise for the whitefish: 'Though it is a rich, fat fish, instead of producing satiety it daily becomes more agreeable to the palate, and I know, from experience, that though deprived of bread and vegetables, one may live wholly upon this fish for months, even years, without tiring.'

It snowed on 20 August and the voyageurs were quite anxious to begin their return—they kept telling Franklin that he had lingered too long. It was far too late and the season far too advanced to return all the way to the Coppermine River, so Franklin brought his canoes to the mouth of what would be called the Hood River (after Robert Hood), destroyed all but one of the birch-bark vessels, in order to lighten the load, and then struck out over barren, rocky terrain inauspiciously known as the Bad Lands towards Fort Enterprise.

But now everything began to go wrong. The deer, which had been plentiful, began to head south as the winter began in earnest. The expedition found that they had to cross numerous small rivers, an arduous process. The perilous nature of their journey was not lost on Richardson that September. 'If anyone were to break a limb here,' he wrote in his journal, 'their fate would have been melancholy indeed, we neither could have remained with him, nor carried him on with us.' One day, Franklin fainted from hunger. On another, the voyageurs dropped the canoe they were portaging, smashing it to smithereens.

By the end of September the party had no supplies left and they were forced to make do with whatever they could find in the forest. There was little game, and the men began to eat the lichen known as *tripe de roche*. Fortunately, on 26 September, they finally found a river whose size meant it was very probably the Coppermine—they also discovered the rotting carcase of a deer, killed by a wolf, and they devoured it right down to the marrow. However, they needed to cross the river to get to Fort Enterprise and they now had no canoes. The frigid, fast-flowing stream was over 90 metres (100 yards) across. Finally, Richardson volunteered to swim the expanse with a rope. Taking off his clothes—John Franklin was to write in his journal of the shock the party felt 'at beholding the skeleton which the doctor's debilitated frame exhibited'—Richardson entered the river. However, he later wrote, 'I had advanced but a little distance from the bank with a line around my middle when I lost the power of moving my arms through the cold. I was then obliged to turn on my back and had nearly reached the opposite shore, when my legs also became powerless, and I sunk to the bottom.'

The men hauled him back and rolled him up in a blanket, but the skin on the left side of his body remained numb for nearly five months.

'You had better kill and eat me'

After a week of trying and with the loss of one man—a voyageur who wandered off in a fit of madness and was never seen again—the expedition managed to construct a canoe and cross the Coppermine. Now, they estimated, they were about 65 kilometres (40 miles) from Fort Enterprise, where food provided by Akaicho's hunters would be waiting for them. In good weather and at usual strength, the men might make this distance in less than a week. But the weather had turned frigid and windy, and many of the men were at the point of collapse. Robert Hood was a walking skeleton, unable to stomach the *tripe de roche*, which had an acidic quality to it. On 4 October two voyageurs collapsed and Richardson was summoned to attend to them. One of them died almost immediately, and he was unable to locate the body of the other in the drifting snow.

In the meantime, the voyageurs were in open mutiny. They told Franklin that they wanted to drop their heavy loads and strike off directly to Fort Enterprise. Franklin at last agreed to send George Back to Enterprise, along with several voyageurs, in order to bring back food and help from the Indians there. Once this group set off, Richardson told Franklin that Hood was too weak to move and suggested that he leave Hepburn behind with Richardson and Hood so they could nurse the latter back to health. Franklin agreed, not without some misgivings, and, leaving the three men a tent to shelter in, headed off in the tracks of Back and his party.

On 7 October four of the voyageurs accompanying Franklin—Jean Baptiste Belanger, Ignace Perrault, Antonio Fontano and Michel

Teroahaute—said they could no longer travel on and asked Franklin to let them go back to Richardson's camp. Franklin, too tired to protest, let them go. His journal notation for that evening was: 'There was no tripe de roche ... We drank tea and ate some of our shoes for supper.'

On 9 October Michel Teroahaute showed up alone at Richardson's camp. He carried a note from Franklin stating that Teroahaute, along with his companions, had decided to wait with Richardson for help. Teroahaute, part French and part Iroquois Indian, was unable to read and at first made no mention of his companions. But when Richardson pressed him, he said that he had become separated from them and did not know where they were.

Teroahaute's presence, at first, was welcome—he brought with him a partridge and a hare that he had caught that morning—and Richardson did not think to question him too closely about his companions. Richardson even gave the voyageur one of his shirts. But it gradually became clear to Richardson, Hepburn and the ailing Hood that something was amiss with Teroahaute. He insisted he sleep separately from the rest, at some distance out in the woods, and also that they leave him the party's hatchet. And he kept disappearing, making up various excuses for his absences—that he was chasing deer or had left a musket behind somewhere.

On 10 October Teroahaute returned to the group with some pieces of meat that he said came from the corpse of a wolf that had been gored by a deer. Richardson and his fellows ate this gratefully, not at first questioning the story, but over the next few days they became concerned about the voyageur's increasing absences, during which he always took the hatchet, and his threatening and surly manner—at one point he told Richardson: 'There is no use hunting, there are no animals, you had better kill and eat me.' As this was occurring, Hood, unable to stomach the *tripe de roche*, was becoming weaker and weaker. On the morning of Sunday, 20 October, both Hepburn and Richardson were briefly away from the camp. Teroahaute was 'lingering by the fire', cleaning his gun, while Hood read a religious tract.

Suddenly, Richardson heard a shot. Staggering back into camp, he found Hood dead of a gunshot wound, with Teroahaute standing there holding a long rifle. Teroahaute belligerently claimed that the young lieutenant had killed himself, but when Hepburn and Richardson inspected the body, they saw that the bullet wound was in the back of the head, fired from so close the powder had burned the nightcap Hood was wearing.

Hepburn had heard Hood exchanging angry words with Teroahaute just before the shot rang out, and it was apparent to both him and Richardson that Hood's death had been homicide, not suicide. There now followed three days of stark terror during which the voyageur raged at the two Britons, daring them to accuse him of Hood's death. He tried to convince Richardson to go hunting in the woods alone with him, but Richardson firmly refused. Finally, on 23 October, Teroahaute left the two men alone, supposedly to hunt for *tripe de roche*, for the first time since Hood's death. Hepburn and Richardson immediately shared their suspicions—to wit, that Teroahaute had killed his fellow voyageurs, was disappearing with the hatchet to cut flesh off their bodies (a hatchet was needed in the frigid climate to hack meat from animals) and had murdered Hood as well. As far as they were concerned, there was only one thing to do. They had two pistols with them and they loaded them.

Richardson described what happened then: 'Immediately upon Michel's [return] I put an end to his life by shooting him through the head with a pistol,' Richardson wrote. He and Hepburn noted that the Iroquois had no *tripe de roche* with him and had loaded his gun, probably 'with the intention of attacking us'.

'An abode of misery'

Although Teroahaute's death would prove controversial, Richardson and Hepburn were convinced they had taken the only course possible to prevent themselves being murdered. After eating some of Hood's buffalo robe, they pressed on through the wilderness, finally reaching

Fort Enterprise on the evening of 29 October, with Richardson so weak that he collapsed in the snow twenty times trying to cover the last 100 metres (100 yards) to the huts.

Murder most foul?

Although he presents it very simply in his journal ('Immediately upon Michel's coming up, I put an end to his life by shooting him through the head with a pistol'), killing the Iroquois voyageur was a shocking act for an officer in the Royal Navy, particularly because the evidence Richardson presents in support of his actions is, in the main, circumstantial. No one saw Teroahaute murder the other men who had been sent back with him—it is possible they died of exposure or illness. And although Hood's death by rifle shot wound in the back of the head is suspicious, it is not conclusive—accident is remotely possible.

Willard Ferdinand Wentzel, the North West Company clerk who accompanied the expedition and who hired most of the voyageurs on Franklin's behalf, was particularly harsh on Richardson, writing later that he had been guilty of 'an unpardonable want to conduct' and 'richly merited to be punished'. And it is true that extreme conditions such as the ones these men encountered can unhinge minds, and make people prone to paranoia and violence.

Yet certain events that Richardson describes make his actions appear just and reasonable within a survival setting. Whether or not Teroahaute killed the three men he had journeyed with, he was almost certainly disappearing from camp to eat their flesh—in fact, one of the reasons Richardson and Hepburn survived, ironically enough, is that he gave them some of what was probably human flesh to eat. Secondly, there is almost no way to account for Hood's death except as a murder. Hood was too weak to have even been cleaning his gun, as Teroahaute claimed may have happened, and suicide was impossible given the location of the wound. Too, Teroahaute had a motive—Hood was slowing them all down. Finally, his excuse that he was leaving the camp for the last time to gather lichen was transparently false. Although the lichen was present nearby, he had gathered none. More likely, he had gone to eat from his hidden cache of frozen human flesh and then returned, gun loaded, intending to do away with the two British officers.

To their horror, they found Franklin and three voyageurs lying on the floor, starving to death. The hut, Richardson wrote, 'was an abode of misery'. The men's appearance was 'hollow and sepulchral' and they were swollen with the oedema that is a precursor to the final stages of starvation. It turned out that Akaicho had not left any food behind for the party and that Franklin had sent Back and the others out to find the Indians. In the meantime, they were all slowly starving to death. Despite his own emaciation, Richardson was still stronger than Franklin and his men and he attempted to nurse them as best he could, melting water for them, making a poor kind of soup, and lancing their swollen legs to let out fluid. The men stayed alive until 7 November by scraping particles from deer hides left behind from the previous winter, in particular by eating the protein-rich grubs laid by warble-flies in the putrid skins.

On 7 November Indians sent by Back, who had located Akaicho's tribe, arrived with food. They were so shocked by the condition of the men that they began to weep and spent days nursing them back to health. Ultimately, the shrewd Orkneymen had been right about this expedition—only nine out of the twenty men who had set off with Franklin in July survived.

After spending a month recovering his strength, Richardson set out with Franklin, Back and Hepburn for York Factory, from where they returned to Great Britain in the spring of 1822. Before setting sail, Richardson sent ahead a letter of warning to his wife: 'You must be prepared to behold traces of age upon my face that have been impressed since we parted,' he wrote. 'This, however, is the common lot of humanity, and I have only taken a sudden start of you by a few years; hereafter I hope we shall grow old together.'

'An inhabitant of these wilds'

Richardson's fond hope was not to be realised—his first wife, Mary Stiven, was to die in the early 1830s, to be ultimately replaced by two more wives, both named Mary, who also predeceased Richardson.

From 1825 to 1827, Richardson returned to the Canadian Arctic with John Franklin and George Back, completing the gargantuan task of mapping the northern shoreline of Great Bear Lake, the fourth largest lake in North America. Richardson also helped map almost 1500 kilometres (900 miles) of Arctic shoreline between the mouths of the Mackenzie and Coppermine rivers. Returning home to Scotland, he was appointed Physician to the Fleets and became senior surgeon of the Royal Navy Hospital at Haslar. He also published the great work *Fauna Boreali-Americana*, which gained him a reputation as one of the great naturalists of his time. And in 1846 he was knighted.

Still, even with these honours, Richardson made one last trip to the Arctic in 1848, when he was sixty years old. The object was to find what had become of his old companion, John Franklin, whose 1845 expedition to the Arctic had disappeared without a trace. Accompanied by the Scottish surgeon and explorer Dr John Rae, Richardson found his way back to the scenes of his old adventures of some twenty-odd years before. He spent two years there, without finding Franklin. In 1850 he wrote to his daughter from the Canadian Arctic: 'Three and twenty years have passed since papa spent one of his birth-days [here] ... It is a long period to look forward to, my dear little child, and to you, I fancy, it looks like a life-time, but I look back on twenty-five years as a dream of the night ... I almost fancy at times that I have never been anything but an inhabitant of these wilds.'

Frightened by persistent chest pains, Richardson left Rae to continue the search for Franklin and returned to Britain. He was to live until 1865. George Back, the last of the 1821 expedition members to survive, wrote of him: 'He was ever a pleasant companion, and, better than all, a moral, good man.'

DAVID DOUGLAS
Scientific traveller

The Scottish botanist David Douglas was so ornery a human being that his mother sent him to school when he was only three years old, just to get him out of the house. His journal is filled with acerbic descriptions of people and places and things that failed his demanding standards—commenting on the cleanliness (or lack thereof) of the vessel that took him to America, he wrote: 'I couldn't help but observe how dogs eagerly licked the decks.' And his sharp eye punctured even a growing young nation's self-confidence: 'I learned of a fine plum named *Washington*, a name which every product in the United States that is great or good is called.'

Yet at the same time, Douglas—who walked thousands of kilometres through the North American wilderness, once trekking from the Pacific coast to Hudson Bay—was a man who exacted the same high standards of himself, whether he was fighting off Indians and grizzly bears or collecting and discovering over two hundred species of plants and trees. His sufferings in the pursuit of knowledge—sufferings that would leave him almost blind and broken in health—seem almost biblical in nature. And behind the astringency of his personality was, those who got close to him knew, a man who loved people dearly but could not show it. A fellow botanist called Douglas 'the shyest being almost that I ever saw'. Although Douglas never married or had children, he was deeply mourned upon his premature, tragic and somewhat mysterious death in Hawaii when he was only thirty-five. Soon after that, however, his discoveries would spread literally over the world. Today, when we smile upon the variety of flowering plants in our gardens, we have David Douglas to thank.

'A devil better than a dolt'

David Douglas was born into the ancient Douglas clan in the town of Scone, Scotland, in 1799. His father was a stonemason and David was the second of six children in a family that had enough food to eat and clothes to wear, but not a great deal more. David was so fractious and obstinate that his mother despaired of caring for him and sent him to school at the age of three, two years ahead of most children of the time. Douglas would have an uneasy relationship with schoolmasters and education, however. Naturally bright and certainly hardy enough—he was a slight youth but by the age of seven he was walking 10 kilometres (6 miles) to and from school every day—he nonetheless 'disliked the restraint of school', as his older brother John was to write, and would skip school again and again in order to fish or wander the nearby fields studying plants, about which he was passionate. As a result, his brother said, 'he was often punished at school'.

It is no wonder that David could not stay in school past the age of ten. He was fortunate, however, in that his father was stonemason to the earl of Mansfield, whose Scone Castle dominated the region. There was room for an apprentice gardener at Scone, and David was given the job. Finding an occupation that he loved, he applied himself to it, in the process impressing the head gardener, Willie Beattie, so much that Beattie took the time to teach him the rudiments of botany. (Douglas was still headstrong and stubborn, but Beattie declared that he preferred 'a devil better than a dolt'.) After serving seven years' apprenticeship at Scone, the now nineteen-year-old left to work at another grand estate some 40 kilometres (25 miles) away—Vallyfield, belonging to Sir Robert Preston. His work there was so impressive that he was appointed to a post in the Botanical Gardens at Glasgow, where he became pupil and friend to the noted naturalist William Hooker. Together the two of them roamed the Highlands, collecting and classifying plants of all kinds. Hooker was quite impressed with Douglas's uncomplaining acceptance of hardships of all kinds—the Highlands could be a very tough collecting environment—and wrote after Douglas's death that his student's 'great activity, undaunted courage, singular abstemiousness and energetic zeal at once pointed him out as an individual eminently calculated to do himself credit as a scientific traveller'.

Douglas's rapid rise continued in 1823 when Hooker recommended that he become a scientific collector for the prestigious Horticultural Society in London. Founded in 1804, the society's prestigious charter members included Sir Joseph Banks (see page 79) and William Forsyth, head gardener to King George III, after whom forsythia is named. The society was employed in sending out 'collectors' around the world to find new plants and rare specimens—a job not without its perils. Of three collectors sent out to India, Africa and China in the year before Douglas's employment began, two died of rare fevers.

Douglas, it was hoped, would not have that problem, since he was being sent to the rather more civilised east coast of America on a relatively tame mission—to bring back fruit trees for British cultivation, especially the hardy American apple tree, as well as to find new strains of oak trees, as the English oak had been greatly diminished because of its use in building warships.

Douglas longed to find the true, wild and unexplored North America, but New York would have to do for now, and thus he set sail for that city in June of 1823.

'My first day in America'

After a difficult two month Atlantic crossing in 'dull, heavy weather', Douglas arrived in New York, so happy to be off the boat that he did not even mind when he was instructed to buy new clothes, since American customs officials were afraid English travellers might bring smallpox into the country with their belongings. Once ashore, he showed letters of introduction to two American members of the Horticultural Society, a Mr Hogg and a Dr Hosack, and they helped him get his bearings and begin exploring. He was delighted with the oaks and maples that grew wild on Staten Island and wrote approvingly about an early morning visit to New York City's Fulton vegetable market where he remarked on 'the immense supply of melons and cucumbers ... and an abundant supply of early apples, pears, and peaches ... [although] the peaches looked rather bad'.

After a few days, Douglas and Mr Hogg headed for Philadelphia, a city he found far cleaner and more inviting than New York—where hogs were used to clean up garbage in the street. He made a point of viewing the specimens the explorers Lewis and Clark had brought back from their 1803–06 expedition across North America. Outside the city, he toured 'the place of the venerable John Bartram', a fabulous botanical garden Bartram, a Quaker, had planted along the Schuylkill River (see page 66).

At the end of August, Douglas left Philadelphia and returned to New York, but in early September he departed again, taking a steamboat up the Hudson River towards Albany and from there, travelling overland and by lake steamer through the towns of upstate New York to the western shores of Lake Erie and up the Detroit River. It was only when he got to these relatively wild areas that he found himself doing what he really wanted to do—not visiting botanical gardens, but foraging through the woods. Impressed by the blazing reds and golds of the American autumn foliage, Douglas wrote happily on 16 September, after a congenial host had taken him out into the woods near Amherstburg, on the Canadian–American border:

This is what I might term my first day in America. The trees in the woods were of astonishing magnitude. The soil, in general, over which we passed was a very rich black earth, and seemed to be formed of decomposing vegetables ... Towards mid-afternoon, the rain fell in torrents, urging us to leave the woods drenched in wet.

Only a collector like David Douglas would consider this day a signal day, but it was the beginning of his love affair with the woods and plant life of North American forests, a love affair that continued unbroken, even when a guide he hired a few days later grabbed Douglas's coat 'and made off as fast as he could run with it'. The coat contained a typical botanist's treasure—a copy of Persoon's *Synopsis Plantarum* and Douglas's pocket vasculum, or sample container. Douglas gave chase, leaping and bounding through the woods, but could not catch the man.

To the Pacific Northwest

In January of 1824 Douglas returned to London, taking with him numerous plant samples, seeds, cuttings and small fruit trees. The Horticultural Society called his trip 'a success beyond expectation', impressed with the quality of his collection, its diversity (twenty-one different varieties of peaches) and also by the fact that Douglas travelled cheaply and kept his nose to the grindstone. Thus, they approved the next trip he requested, to an area that was fast becoming one of the most exciting new territories for exploration in the world—the Pacific Northwest. Captain George Vancouver had landed there in 1791 (with naturalist Alexander Menzies aboard), followed by the Scottish-Canadian explorer Alexander Mackenzie in 1793. Lewis and Clark, after travelling the breadth of the American continent, had arrived at roughly the same spot—the mouth of the Columbia River in the present-day state of Oregon—in 1805.

The Hudson's Bay Company was busy building a fur empire in the Pacific Northwest but as yet no full-time naturalist had explored the area, and the Horticultural Society concurred with David Douglas that he should be the first one. The cost-conscious society prevailed upon the Hudson's Bay Company to allow Douglas to sail on its annual supply ship heading to the mouth of the Columbia River (kindly telling Douglas that the fare on his journey would be 'rather coarse and subject to some privations'). In July of 1824, he set off on his second voyage to North America, this time making the classic journey down the Atlantic, around the Horn, then up the east coast of the Americas. Along the way, he stopped at Más a Tierra and also made landfall in the Galapagos, well before Charles Darwin ever got there. Douglas was duly impressed by the variety and profusion of wildlife there—with birds simply alighting on the barrel of his gun, so unaccustomed were they to humans—but frustrated that he was allowed only a short time for specimen-collecting.

Finally, on 9 April 1825, Douglas arrived at Cape Disappointment, after weathering a storm that kept the ship offshore for six weeks, unable to enter the Columbia River, the fog-enshrouded coastline tantalisingly in sight. When Douglas came staggering ashore, he was pleased to find 'plants only known to us through the herbaria'—such as *Rubus spectabilis* and *Gaultheria shallon*—and immediately set about picking and preserving them. Douglas carried a letter of introduction to John McLoughlin, chief factor of the Hudson's Bay Company on the Columbia, and McLoughlin arranged to send him by canoe 110 kilometres (70 miles) up the river to Fort Vancouver, where Douglas began his studies of North American plant life in earnest.

'Pines at my pleasure'

Wandering now along the Columbia and its tributaries, Douglas was in his element. In his journal, he rhapsodised over the landscape:

> *The scenery from this place [Fort Vancouver] is sublime—high well-wooded hills, mountains covered with perpetual snow, extensive natural meadows and plains of deep, fertile alluvial deposit covered*

with a rich sward of grass and a profusion of flowering plants. The
most remarkable mountains are Mounts Hood, St. Helens, Vancouver
and Jefferson, which are at all seasons covered with snow as low down as
the summit of the hills by which they are surrounded.

It was during the early part of this journey to America that Douglas walked into a strand of evergreens that were wonderfully straight and tall, taller than any that he had yet seen. They were the marvellous trees that would be known to posterity as Douglas firs and they were but the first of the evergreens that Douglas would discover and name in America—the Sitka spruce and the sugar pine would be next. With great delight, Douglas wrote to his mentor, William Hooker: 'You will begin to think that I manufacture Pines at my pleasure.'

Travelling with an Indian guide—although often forging ahead on his own, in his impatient way—Douglas gradually learned the customs of the region. Following a rough day in the wilderness, he wrote, he was only too glad of 'fresh salmon, without salt, pepper or any other spice, with a very little biscuit or tea, which is a great luxury after a days march'. (Douglas was astonished at the quality and quantity of the salmon during the spring salmon runs. He watched as Indians took 1500 at a time, each weighing almost 7 kilograms (15 pounds), using either bone spears whose tips were attached with ropes, or small nets or hoops.) He soon gave up any notion of hanging onto such niceties of civilisation as bedrolls and pillows: 'The luxury of a night's sleep on a bed of pine branches can only be appreciated by those who have experienced a route over a barren plain scorched by the sun or fatigued by groping their way through a thick forest ...'

He was, in other words, in his element. In less than six months he travelled more than 3000 kilometres (2000 miles) and collected almost four hundred specimens, including the *Helianthus* (sunflower), the fleabane, the wild hyacinth, the evening primrose and numerous other plants.

After wintering in Fort Vancouver, where he spent his time categorising and packing his collection to be sent to Britain, he set off again in the spring of 1826, tagging along with a fur-trading expedition to the Spokane River and the grand Kettle Falls. It was during this time that he had one of his first encounters with warlike Indians. A group surrounded the expedition and took the usual present of tobacco but then refused to allow them to leave. Douglas observed some of the Indians throwing water on the gunlocks of the expedition's rifles—so they would not fire—and warily reached for his gun. When one of the Indians fitted an arrow to a bow and made as if to threaten one of the fur traders, the botanist acted quickly: 'I instantly clipped the cover off my gun, which at this time was charged with buckshot, and presented it at [the Indian], and invited him to fire his arrow, and then I should certainly shoot him.'

The Indians backed down and the legend of David Douglas in the wilderness grew.

'A remarkable large pine'

On this same journey, Douglas entered the Blue Mountains (which extend from northeastern Oregon to southeastern Washington state). He spied a mountain he wanted to climb one day in June, but his Indian guide refused to go with him. Typically, this did nothing to deter Douglas. As he wrote:

> *In the lower parts [of the mountain] I found it exceedingly fatiguing walking on the soft snow, having no snowshoes, but on reaching within a few hundred feet [100 metres] of the top, where there was a hard crust of snow, I without the least difficulty placed my foot on the highest peak of these untrodden regions where never European was before me ... The view of the surrounding country is extensive and grand. I had not been there above three-quarters of an hour when the upper part of the mountain was suddenly enveloped in dense black cloud; then there commenced a most dreadful storm of thunder, lightening, hail and wind ...*

Finally feeling the need for caution, Douglas made his way down the mountain in near darkness and found his way to camp, but nearly froze to death in wet clothes, since no fire could be made. The next day he could hardly walk but, of course, he pressed ahead anyway.

The previous year, Douglas had become curious about the seeds Indians kept in their pouches and ate as snacks—these, he believed, were the seeds of a 'remarkable large pine' he called *Pinus lambertiana*, and he searched for it everywhere. By showing Indians he met the seeds and questioning them closely, he finally understood that the tree existed near the Umpqua River in central Oregon. As he entered the area that October, a fierce storm blew in and trees came crashing down all around his camp. But the next day Douglas came upon a tree blown down by the storm and realised that 'it was my long-wished *Pinus*'. He stared at it in awe, later writing in his journal:

> *New or strange things seldom fail to make great impressions, and lest I should never see my friends to tell them verbally of this most beautiful and immensely large tree, I now state the dimensions of the largest one I could find that was blown down by the wind: 57 feet, nine inches [17.6 metres] in circumference ... extreme length, 215 feet [65.5 metres].*

For Douglas, it was the Holy Grail. Of course, this grand find was immediately followed by danger. Attempting to shoot down some 45-centimetre (18-inch) pine cones hanging 'like small sugar loaves in a grocer's shop' from one of the living trees (in order to take them as samples) he attracted the attention of a roving band of Indians, one of whom threateningly began to sharpen his knife right in front of Douglas. Douglas cocked his gun and also reached for one of his pistols. He was outnumbered, but the bluff worked and the Indians backed away. He then told them he would give them tobacco if they collected pine cones for him, and when they went off, he fled back to his camp, where he spent a restless night whose discomfort was further increased, early the next morning, by the arrival of a grizzly bear and her two cubs. Astonishingly, Douglas merely mounted his horse, walked it to within 20 metres (20 yards) of the bears, shot one of the cubs in the face and then 'lodged a ball' in the mother's chest.

'One of the most striking objects in nature'

While David Douglas was not the first European to discover the Douglas fir—that honour goes to his countryman, surgeon and naturalist Archibald Menzies, who first noted the tree on Vancouver Island in 1791—he was the one who named it and introduced it to Europe.

The Douglas fir is a tall, graceful and hardy evergreen that grows on the eastern and western side of mountains all the way up western America, from southern California to Vancouver, Canada. The hardy Douglas is drought-resistant, grows well in burnt-out areas and also does well in shady areas.

David Douglas found one Douglas fir that he measured at 69 metres (227 feet) tall and 14.6 metres (48 feet) in circumference, and there are even taller ones—the tallest to be found currently is 102.5 metres (336 feet) high; one Douglas fir cut down in the early part of the twentieth century measured as high 126.5 metres (415 feet). And the trees are amazingly long lived. There are Douglas firs alive today that David Douglas himself set eyes on; one Douglas fir, in Vancouver, Canada, is thought to be 1300 years old.

The main reason for the popularity of the Douglas fir, of course, is how useful it is. Douglas wrote that the fir was 'one of the most striking objects in nature' and predicted that it 'will prove a beautiful acquisition … the wood may be found very useful'. In 1827 he introduced the tree to Great Britain, which had heretofore only three species of conifer. It grew easily in the cool, wet British climate. Its wood is supple and soft, yet strong, and it is used in house building, cabinetmaking, wood panelling and flooring.

'Smells, smoke and grime'

New species, mountains, Indians and grizzly bears were seemingly all in a day's work for Douglas. In the spring of 1827, he left Fort Vancouver and headed overland to Hudson Bay, walking thousands of kilometres. At one point, as he and his party were nearly starving, one of the Indian hunters shot a partridge—and Douglas, seeing it was a new species, insisted on preserving it for his specimen bag. He arrived at York Factory, the Hudson's Bay Company's headquarters in Hudson Bay, after a journey of some six months, but he still had the strength to pick a fight with a boorish trapper (who then challenged Douglas to a duel, an invitation Douglas simply ignored).

From Hudson Bay, Douglas sailed across the Atlantic, arriving in England in October of 1827, where he was soon quite famous for his achievements. However, he was battered. His knees were shot, his body wracked with fever, and he was beginning to experience bouts of cloudy vision, possibly from an infection or prolonged exposure to the sun. After such a long time out of civilisation, he became depressed and even more irascible, and soon, as his biographer George Harvey has written, 'became disgusted with the smells, smoke and grime' of London. He began to lash out at his friends—when one gave a lecture, Douglas wrote grimly, 'the beginning was bad, the end was bad and the middle worthy of the beginning and the end'.

Part of this anger may have come from the fact that, in some ways, Douglas did not belong. Despite his new and hard-won fame, he was still the poorly educated son of a stonemason, and some of his plant classifications were already being altered by more scientifically trained naturalists. The only solution for it all—for both the Horticultural Society and Douglas—was for Douglas to head back out into the field. This he did, with a vengeance. Starting in 1829, he travelled again to the Pacific Northwest, spent most of the year there and then headed south to explore California, visiting redwood forests north of San Francisco. He spent all of 1831 and 1832 in California, sending back hundreds of specimens and becoming friendly near Monterey with Franciscan monks, who admired his devotion to nature. In 1833 the restless Douglas journeyed back up to the Columbia River, to Fort Vancouver, and in the spring he attempted the grandest journey of all. He decided to head north through British Columbia to Alaska and from thence across the Bering Strait to Russia, returning to Britain via Europe.

It was not to be. By this time his eyes were bothering him badly. He was nearly blind in one eye and the other was so light-sensitive that he was forced to wear violet glasses, which, he sadly observed, rendered 'every object, plants and all ... the same colour'. The Alaska trip was so arduous that he was forced to turn back before even getting there; unfortunately, on the way back to Fort Vancouver his canoe capsized and all four hundred of the specimens he had gathered were lost.

Douglas decided to set sail for Hawaii in October 1833, thinking the tranquil weather and a period of rest would do him good. From there, in 1834, he would sail back to Britain and an uncertain future.

The wild bull pit

Douglas arrived on Hawaii Island at the end of December 1833, determined to rest and work on his specimens before return to Britain. But the lure of the wild was too much, especially the spectacle of Hawaii's volcano Mauna Kea, which rose 4205 metres (almost 14,000 feet) into the sky. Arriving near the volcano in January, he stayed with the Reverend Joseph Goodrich in the town of Hilo. Aided by Goodrich—who helped him find guides—Douglas began his ascent. It was a typical Douglas saga. 'We continued on our way,' he wrote, 'under such heavy rain, as, with the already bad state of the path, rendered walking very difficult and laborious; in the chinks of the lava, the mud was so wet that we repeatedly sunk in it, above our knees.'

After several days intense climb, they reached the highest peak of the mountain, Great Peak. The rain had stopped and the heat given off by the hidden lava of the volcano was such that it blistered Douglas's feet when he was foolish enough to walk around the summit without his shoes on. Despite this discomfort and a violent headache, possibly brought on by altitude sickness, Douglas drank in the glorious view, collected samples of mosses and lichens, and then headed back down the mountain.

By the end of the month, Douglas had ascended another great Hawaiian volcano, Mauna Loa, but his health was worsening. His eyesight continued to deteriorate; at one point, his eyes literally bled. Even so, he continued. After returning to Hilo, he met another missionary, an Englishman named John Diell, and they decided that they would climb another volcano, Mauna Kilauea. As they travelled overland, Diell decided to make a side trip to the island of Molokai, but Douglas journeyed onward. On 12 July, accompanied only by his dog, Billy, Douglas met a wild cattle hunter by the name of Ned Gurney. Gurney was a shady character, an English thief who had been transported to Australia for his crime and who had escaped (or possibly served his time and been released) and found his way to Hawaii.

According to Gurney's story, he had warned the nearly blind naturalist about the presence of pits he had dug to catch wild bulls. Douglas had accepted this information and asked Gurney to guide him a little way along the trail, past one of the pits. At this point, Gurney left Douglas. It appears that, for some reason, Douglas retraced his steps to the wild bull pit, where he fell in and was trampled and gored by a trapped bull. Shortly afterwards, two Hawaiians found the body in the pit and informed Gurney, who had it transported to Hilo.

At least this is the official story—there were numerous suspicions surrounding Douglas's death. However he died, the fact remained that one of the most determined and brilliant naturalists of his time had perished at the age of thirty-five. It is true that Douglas was not a scientifically trained observer and that many of his classifications were later found to be incorrect. But Douglas, a great traveller and collector, had far more impact on the world than many of the now-forgotten scientists of his time. Because of Douglas, the Douglas fir and other major trees have spread around the world and his American flowers now sprout in gardens throughout the world.

Returning from one of his Hawaiian volcano climbs, Douglas wrote in his journal: 'It is with thankfulness that we approach a climate more congenial to our natures and welcome the habitations of our fellow man.' Although ultimately an isolated man, Douglas ended his life seeking connection.

'Our hearts almost fail us'

Dated 15 July 1834, the letter was addressed to Richard Charlton, the British consul in the Hawaiian Islands. It began:

> DEAR SIR—Our hearts almost fail us when we undertake to perform the melancholic duty which devolves upon us, to communicate the painful intelligence of the death of our friend Mr. Douglas, and such particulars as we have been able to gather respecting this distressing providence.

The missive was from missionaries John Diell and Joseph Goodrich, who had befriended David Douglas on his arrival in the islands. It was to these men, in the village of Hilo, to whom Ned Gurney had sent Douglas's battered body after his death in the wild bull pit. Gurney had paid a party of Hawaiians four bullocks to transport the corpse nearly 50 kilometres (30 miles) to Hilo. The missionaries received it with great distress. They describe the condition of the corpse in their letter: 'The face was covered with dirt, the hair filled with blood and dust; the coat, pantaloons, and shorts considerably torn ... On washing the corpse, we found it in a shocking state: there were ten to twelve gashes on the head ... The left cheek appeared to be broken, and also the ribs on the right side. The abdomen was also much bruised and also the lower part of the legs.'

The missionaries were preparing to bury Douglas—having accepted the story sent along by Gurney, that the botanist had fallen into the bull pit accidentally—when Charles Hall, an American wild bull hunter who happened to be present and who was aiding Diell and Goodrich, stated that he felt the gashes on Douglas's head were inconsistent with those that might be inflicted by a bull's horns, especially when the bull in question was reportedly old, with relatively short and blunted horns. The missionaries then decided to preserve the body by filling its stomach cavity with salt and send it to Honolulu for further study. However, by the time the corpse reached Consul Charlton in Honolulu, it was severely decayed. Several British naval doctors quickly examined it, but their finding was one of accidental death.

Still, rumours persist even to this day. In the main, they are based on the unsavoury reputation of Gurney and the theory—never really proven—that Douglas was carrying a large sum of money on his person in order to pay for his voyage back home. The way this theory goes, Gurney murdered Douglas with an axe and then threw his body into a pit with the bull. There is some circumstantial evidence to support this. Douglas's purse was missing and only a few dollars were found on him; he would indeed have needed a much larger sum to pay for his transport back to Britain. There were those who believed that Gurney's payment of four bulls to the Hawaiians—a hefty price to transport the body of a person Gurney had just met that morning—was suspicious. Too, many people found it impossible to believe that such an experienced botanist, even one with failing eyesight, would fall into a large pit containing a dangerous wild beast, particularly when he had been warned about it.

Others, however, pointed out that even experienced bull hunters occasionally fell into the pits. And, seven years after Douglas's death, two English scientists visited the area and noted that most of the bull pits were disguised with earth and 'fragile plants' such as raspberries, which leaves one with the tantalisingly ironic notion that Douglas died as he lived, reaching for and plucking native fruit.

MARY ANNING
Fossils in the Blue Lias

She sells seashells on the seashore
The shells she sells are seashells, I'm sure
So if she sells seashells on the seashore
Then I'm sure she sells seashore shells.

Almost everyone was taught this popular tongue-twister as a child—along with 'Peter Piper picked a peck of pickled peppers', among others—but very few people realise that the author of the ditty, Terry Sullivan, writing in the late 1890s, was referring to a very specific 'seashell seller'. The subject of his ditty was a woman named Mary Anning who roamed the lime and shale cliffs of the stunningly beautiful coast of southern England in the first half of the nineteenth century. Only it was not really seashells she found and sold, but fossils that were millions of years old. At a time when human understanding of the past was for the most part encapsulated in the story of Genesis, where God built the world in six days and on the seventh rested, this young woman from a poor hard-luck family would help change our vision of ourselves forever.

Lyme Regis

The town of Lyme Regis, in Dorset, on the southern coast of England, is steeped in history. It dates to the eleventh century, when it was known in the Domesday Book as Lyme—in the thirteenth century a Royal Charter from King Edward I added 'Regis' to its name. From the sixteenth century until the late eighteenth it was one of the most important ports in England. Situated only 34 kilometres (21 miles) across the English Channel from France (which can be easily seen on a clear day) it was filled with shipyards and was at one point even larger than Liverpool. The town was known for The Cobb, a breakwater built in the fourteenth century to protect the harbour against the tempests of the Channel; The Cobb, like so much in Lyme Regis, features in fiction from Jane Austen's 1818 novel *Persuasion* to John Fowles's 1969 classic *The French Lieutenant's Woman*. The town is also known for its spectacular surrounding cliffs. Towering 60 metres (200 feet) into the air, they are some 200 million years old, made of alternating layers of limestone and shale, a geological formation known as Blue Lias, whose striking blue-grey colour is beautiful to look at. But the cliffs can be unstable, given to 'landslips' in which large portions of the cliff walls come tumbling down to the beach and ocean.

By the late eighteenth century Lyme Regis's days as a commercial powerhouse were numbered because its harbour was too small and shallow for the large, deeper-draft merchant ships being built. Moreover, the town had always had a large population of weavers—literally a cottage industry—whose cloth was shipped to the rest of England, but by the 1790s the Industrial Revolution had centred cheap cloth production in factories farther to the north in England. Gradually, Lyme Regis began to shrink—and as the war with France heated up at the end of the eighteenth century its closeness to France became a liability, since it was seen as the most obvious spot for the French to invade England. In 1799, when Mary Anning was born there, the town still hosted a fairly sizeable population of 1200 people, but most of them were dirt poor, struggling to make ends meet.

The marked child

Mary was born to Richard and Molly Anning, a couple who had moved to Lyme Regis in 1793 from a small village farther inland. Richard was a carpenter and cabinetmaker who set up shop in Lyme Regis, hoping to take advantage of a tourist trade that was gradually building on the coast now that sea bathing had started to become fashionable. But Richard was not exactly a nose-to-grindstone sort of tradesman. He was a Nonconformist, a practising Dissenter (a religious group that would eventually evolve into the Congregationalists), and a man who did not kowtow to anyone. This included his hoped-for clientele, tourists such as Jane Austen. In 1804, while in Lyme Regis on a visit, Austen took a small box to him, asking him to repair its broken lid. Richard Anning told her he would charge five shillings for the job, a price, as she wrote to her sister Cassandra, 'that was beyond the value of all the furniture in the room together'.

While Richard was busy losing Jane Austen's business, Molly was attempting to raise a family (she would bear nine children) but hard and tragic luck attended this business. By 1798 she had five children, the oldest being a daughter, Mary, born in 1794. On a freezing cold day in December 1798, Molly left Mary and her two-year old brother Joseph alone in a room with a fire burning on the hearth for the space of perhaps fifteen minutes. Hearing horrendous screams, Molly raced back into the room to find Mary a flaming human pyre. She beat the flames out but, as a newspaper account at the time read, 'the girl ... was so dreadfully burnt as to cause her death'.

The following May, Molly had another little girl, whom she named Mary after her dead child. It seemed for a time that this Mary might not survive either, since she was a sickly child, given to croup and colds. But then something occurred that many people considered a miracle. At the age of fifteen months, in August 1800, Mary was taken to view a travelling company of trick riders who were performing in a field on the outskirts of Lyme Regis. Her mother stayed home in order to rest—Mary's caretaker was a friend of Molly's, a nursemaid named Elizabeth Haskings. As the trick riders performed to the delight of hundreds of townspeople who had very little in the way of entertainment in their lives, a thunderstorm boomed overhead. Suddenly there was a horrible crack, a flash of lightning and an explosion. Looking up to a little knoll on which stood a large elm tree, the crowd saw that several women who had taken shelter there were lying still upon the ground.

People raced up the hill to see a horrifying sight. Underneath the elm, which was shattering by lightning, three women lay dead. One of them was Elizabeth Haskings, whose entire right side was charred and blackened, and in her arms was Mary Anning. To the astonishment of the crowd, the infant was breathing. She was taken home, treated by a local doctor (who dunked her in warm water) and within hours seemed to have recovered completely. From then on, for better or for worse, Mary Anning was a marked child.

'Having been dug up'

Even before Mary's triumph over lightning, her father had forsaken his carpentry business for a hobby that, over the years, had built up to a passion—collecting fossils along the cliffs and seashores. Fossils— the word comes from the Latin for 'having been dug up'—were and are plentiful on Lyme Regis's beaches and along what is now known as England's Jurassic Coast, a stretch of Devon and Dorset nearly 160 kilometres (100 miles) long where rock strata represent 200 million years of geological history. Some of the fossils, in Mary Anning's time, lay plentifully on the beach. Others had been frozen for eons into rocks or the bases of the cliffs and were exposed only at low tide. Every powerful storm revealed new ones and destroyed familiar ones.

For years, people had been coming back from the beach with small ammonites, the coiled sea creature whose name derives from its ram's horn appearance; brachiopods, of the Palaeozoic era; coprolites, or petrified animal dung; and many other types of fossils. However, most people in the early part of the nineteenth century did not believe that these were the remains of animals millions of years old, in part because they did not believe that any creature in God's universe actually became extinct—surely these imprints of animals merely indicated that the creature was alive, in some remote part of the earth. Some fossils were particularly difficult to explain—monstrous heads seemingly carved into rock walls (petrified crocodiles, it was said) and pieces of giant vertebrae, known as verteberries (some thought these last were the backbones of dragons). Many people simply thought that fossils made nice adornments. They were given names like ladies fingers, cupid wings and devil's toes and were used to adorn pottery, mirrors and bracelets.

However, these attitudes were gradually changing. An English canal engineer, William Smith, noticed while his canals were being dug that each separate stratum of rock always displayed the same type of fossil; geological time could then be read in the fossil levels. (Smith, unrecognised until 1830 by the scientific establishment, would in 1815 create the first geological map of England and Wales.) The French naturalist Georges Cuvier gave a lecture in 1796 in which he asserted that some species, who 'lived in a world previous to ours' did indeed vanish from the earth, although he attributed this not to natural selection, a theory later postulated by Charles Darwin, but to 'some kind of catastrophe'.

More and more gentleman naturalists had begun to flock to Lyme Regis and its environs, and Mary's father, Richard, was able to show them where to go and to provide them—along with the numerous summer tourists—with specimens. He spent most of his days scrambling over the cliffs with a canvas sack and a hammer, returning late in the evening with bags full of fossils that he would sort out on his workbench, wash and attempt to painstakingly chisel from their stones.

From about the age of six, Mary accompanied her father and her older brother Joseph (they would be the only two of the Anning children to survive out of nine, the other seven killed by childhood illnesses or accidents). It was dangerous work, with the chance of cave-ins and with fierce storms blowing in from the southwest, but far more exciting than going to school. But in 1807 Richard fell over a cliff, a fall that nearly killed him. He gradually recovered but was never able to fossil hunt again, and he died, perhaps of tuberculosis, three years later at the age of forty-four. With no other way to make a living for her family—and with Molly nearly paralysed by depression—Mary realised that she needed to turn fossil-hunting into her full-time profession.

Mary Anning in literature

There is something intriguing and enigmatic about Mary Anning—her solitary, still somewhat mysterious character, her stubbornness, or her obsessive nature—that has called to writers from the mid-nineteenth century on. Charles Dickens eulogised her in his weekly literary magazine All the Year Round, *acknowledging the challenges she faced and writing of her 'high degree of that sort of intuition' without which being a collector would be impossible. He also attacked those in Lyme Regis who valued her merely as 'bait for tourists' and who did not lift a finger to help her as she died painfully of breast cancer. The theme of ill-usage is also struck by John Fowles in his postmodernist masterpiece* The French Lieutenant's Woman *(1969), which takes place in Lyme Regis where Fowles lived from 1968 until his death in 2005. 'One of the meanest disgraces of British paleontology,' Fowles writes in the novel, 'is that though many scientists of the day gratefully used her finds to establish their reputation, not one native type bears the name anningii'.*

Among other books that feature Anning are The Dragon on the Cliff: A Novel Based on the Life of Mary Anning, *by Sheila Cole;* Stone Bone Girl: The Story of Mary Anning, *by Laurence Anholt;* Remarkable Creatures *by Tracy Chevalier; and* The Fossil Hunter; Dinosaurs, Evolution and the Woman Whose Discoveries Changed the World *by Shelley Emling.*

'Fish-lizard'

At first, clambering over the dark cliffs, Mary was joined by her brother Joseph. One day in the spring of 1811, when he was fourteen and Mary twelve, Joseph went on a solitary jaunt to the beach and spied, among the rocks along the base of a shale cliff, a huge skull encased in rock. It looked to Joseph like a dragon or crocodile's head, 1.2 metres (4 feet) long, with two hundred teeth. He brought his find to Mary as a curiosity but thought little else about it, since much of his time was taken up by his apprenticeship to a local upholsterer.

Mary, however, was fascinated by the creature. It took nearly a year for her to find the rest of the fossil, and only then because a huge storm in the late winter of 1812 revealed the first parts of the animal's skeleton petrified in the cliff above the beach about a kilometre and a half (a mile) from Lyme Regis. She took her hammer and began to painstakingly chip away at what were apparently the creature's vertebrae; shortly after this, she found a few ribs deep in the wall of the cliff.

Finally, after several months spent separating it from the surrounding rock, the outline of the creature became apparent. It was remarkable to say the least—a type of lizard, it seemed, 5 metres (17 feet) long, but with appendages that resembled flippers. Mary's mother, Molly, begged several local quarrymen to help carry the fossil, encased in stone, from the beach to Richard Anning's old workshop, where it became a local attraction and gave Mary her first claim to fame—the untutored village girl who had unearthed such a fantastic phenomenon. No one had ever seen such a skeleton, it was thought. The creature, soon to be dubbed *Ichthyosaurus*, had actually been first described from a few vertebrae found as early as the beginning of the eighteenth century, but Mary was the first person to find an entire connected skeleton.

As with all her forthcoming discoveries, however, Mary was forced to sell *Ichthyosaurus* in order to provide for her family. She had made the find on land that belonged to a well-off landowner, Henry

Hoste Henley, who paid her twenty-three pounds, a large sum that would keep her mother and brother in food and pay their rent for some time, but that she probably had no choice but to accept. Henley allowed the *Ichthyosaurus* to be displayed in a museum in London, where people lined up day after day to view it, staring in amazement at its bony outline and spookily large eye sockets.

In 1817 Henley sold the fossil to the British Museum, for twice what he had paid Mary for it. A debate raged among scientists—was the creature a fish or a lizard? The director of the British Museum named it *Ichthyosaurus*, or 'fish-lizard', trying to have it both ways. In fact, he was close. The *Ichthyosaurus* lived from about 250 million to 90 million years ago; it was a giant marine reptile that hunted fish at speeds of up to 40 kilometres (25 miles) per hour, diving deep into the ocean, where its large eyes helped probe the marine darkness.

'Daily went to strew fresh flowers'

Although Mary Anning could read and write she left behind no journals, few letters, no real self-portrait. We catch glimpses from the writing of others, who describe her as not unattractive, with thick black hair and strong features, but rather unkempt, her fingernails blackened and torn from scrabbling into the rock faces, her skin weathered, her clothes—especially her trademark long grey skirt and shawl—ragged and torn. Charles Dickens, writing about Mary Anning after her death in his literary magazine *All the Year Round*, took her life as an example of 'what humble people might do, if they have just purpose and courage enough, towards promoting the case of science' and further wrote that her providing debating scientists with fossil specimens ('the munitions of war', he called them) was her 'calling' in life.

Mary was far too independent a spirit, however, to see herself simply as a purveyor of petrified bones to squabbling intellectuals. A neighbour gave her her first geology book just after the fourteen-year-old had discovered the entire skeleton of the *Ichthyosaurus*, and she read it obsessively, as she did any scientific writing, not just on geology or fossils but also on animal anatomy, a subject that aided greatly in her ability to identify the shadowy outlines of ancient animals buried deep within cliff walls. Her study of anatomy led her to dissect squid (living ones) that she had found in the waters off Lyme Regis, ignoring it when they squirted bodily fluid all over her. Despite this, she was sentimental. In 1815 a British merchant vessel on its way home from India floundered and sank off the Lyme Regis coast, and out of an estimated 160 people aboard, only four or five survived. One body washed ashore near Lyme Regis was that of a woman, known only as Lady Jackson, who had been returning to England with her husband and three children, who also perished. Lady Jackson was quite beautiful and Mary Anning visited her body repeatedly as it lay, first on the beach and then in a local church, waiting to be claimed. Mary's friend Anna Maria Pinney wrote that Mary 'untangled the seaweed which had attached itself to [Lady Jackson's] long hair ... and daily went to strew fresh flowers' over the body.

Mary was sixteen years old at the time, tough, smart and romantic, a small town girl who had never left her coastline. Instead, men of science came to her.

'All these vast intervals seem annihilated'

One of these men was William Buckland, the son of a minister, who grew up in a village only a few kilometres away from Lyme Regis. Unlike Mary, Buckland had the opportunity to leave the coast and make a great success of himself, getting a degree in theology from Oxford and ultimately becoming a professor of geology. But the eccentric Buckland did not put on airs, coming to see Mary shortly after she had become famous for discovering the *Ichthyosaurus*. He was twice the age of the sixteen-year-old and continued his visits during summer vacations. Buckland's daughter would write in her biography of her father: 'The vacations of his earlier Oxford time were often spent near Lyme Regis. For years afterwards, local gossip preserved traditions of his adventures with that geological celebrity, Mary Ann Anning, in whose company he was to be seen wading up to his knees in search of fossils in the blue lias.'

'What rules the world?'

William Buckland was not only a friend to Mary Anning but an extraordinarily colourful personage in his own right. He was born on 12 March 1784 in Axminster, a village just 10 kilometres (6 miles) from the Dorset coast, not far from Lyme Regis. His father was a minister and Buckland studied theology at Oxford and was ordained an Anglican minister in 1808.

But Buckland had an avocation that gradually grew to take over his life—he loved fossils, loved rocks, loved the cliffs around Lyme Regis. 'They were my geological school', he wrote about the latter. 'They stared me in the face, they wooed me and caressed me, saying at every turn pray, pray, pray to be a geologist.' His prayers were answered; by 1813 he had become a professor of geology at Oxford, renowned in his field, beloved by his students, who found him anything but boring. At one point during a lecture, a student wrote, he:

> ... rushed, skull in hand, at the first undergraduate on the front bench—and shouted, 'What rules the world?' The youth, terrified, threw himself against the next back seat, and answered not a word. He rushed then on me, pointing the hyena full in my face—'What rules the world?' 'Haven't an idea,' I answered. 'The stomach, sir,' he cried.

Buckland was also fond of strutting around the classroom imitating the way he thought dinosaurs walked, looking like 'a flustered hen beside a muddy pond', as the novelist John Fowles was to write. And he was famous for his eating habits, specifically for being willing to eat anything. His meals famously included mice on toast, hedgehog, bluebottle flies and moles. Once, supposedly, he even snatched up the mummified remains of the heart of Louis XIV, preserved in a silver casket at an English castle, and gobbled it down.

Not all was fun and games, however. He coined the term 'coprolite' (from the Greek words for 'dung' and 'stone') after examining the 'bezoar' stones introduced to him by Mary Anning. He was the first to understand how much science could learn from examining such a humble substance as petrified poop. And he also famously discovered, from fossils found near Stonesfield Quarry, the megalosaurus, *which means 'great lizard', a name that foreshadowed the word 'dinosaur', coined by the British scientist Richard Owen in 1842.*

Buckland was not just slumming, either. Mary was intimately familiar with the territory and seemed to have an almost preternatural ability to find hidden fossils. One science writer, John Murray, wrote after hunting on the beach with her: 'I once gladly availed myself of a geological excursion and was not a little surprised at her tact and acumen. A single glance at the edge of a fossil peeping from the Blue Lias revealed to her the nature of the fossil and its name and character were instantly announced.'

Mary astonished Buckland and other men of science who came to visit her. Between the ages of sixteen and twenty-one, she discovered three more complete skeletons of *Ichthyosaurus,* which ranged in size from that of a small fish to that of a whale. Some of these fossils still contained the food the animal had eaten before its death, a fact that particularly struck Buckland. As he wrote in a lecture he gave to Oxford students, the *Ichthyosaurus's* outlined skeleton and ribs 'still surrounded the remains of feeding that were swallowed ten thousand or more than ten thousand times ten thousand years ago, all these vast intervals seem annihilated, come together, disappear, and we are almost brought into as immediate contact with the events of immeasurably distant periods as with the affairs of yesterday'.

Buckland—a minister, after all—was struggling with the idea of the vast age of the world, all brought home to him by the fossils Mary had dug up on the beach, an age that seemed to make it clear that the story of creation in the Book of Genesis could, at best, be only an allegorical one. Certainly Mary thought this to be so, and she argued vociferously with Buckland about it. One scientist who visited her reported that 'her account of her disputes with Buckland, whose anatomical science she holds in contempt, was quite amusing'.

The 'proprietor'

In the years to come, Mary would find numerous species of dinosaurs unknown to science. In 1823, following yet another winter storm, she discovered the skeleton of a plesiosaur, a long-necked aquatic lizard, the first ever seen. It caused a huge uproar. The duke of Buckingham paid Mary the huge sum of 110 pounds for the creature, which was eventually shipped to London. William Daniel Conybeare, a wealthy young minister with a geological avocation who was a friend of William Buckland and was as excited by the discovery as Buckland at Oxford, was chosen to give a presentation about the creature to the prestigious Geographical Society. The ship carrying the fossil was delayed in the English Channel and so Conybeare gave his presentation using Mary's own sketches, which were finely rendered and accurately detailed.

Interestingly, though, he made no mention of Mary's role as the discoverer of the plesiosaur. As one of Mary Anning's recent biographers, Shelley Emling, points out: 'As [the speech went on] whenever Conybeare was forced to refer to Mary, he generally used the word 'proprietor,' a somewhat negative description that inadvertently suggested that money was the only motive for her preoccupation with fossils.'

Of course, Mary was forced to sell her fossil finds in order to support herself and her mother, but this did not negate her brilliance and her love of science. But because she was a woman, and because she was poor, she was considered of little consequence by many of the men she was dealing with. Yet she continued to hunt for fossils, day after day, finding bones that belonged to *Hybodus*, the prehistoric predecessor of the great white shark. She also caused Buckland to take another look at the so-called 'bezoar' stones that were often found inside the stomachs and intestinal areas of the fossilised animals; it was Mary who broke these open to find small fish bones inside them, leading Buckland to understand that they were fossilised faeces, and to name them coprolites.

'I am well-known throughout the whole of Europe'

At one point around 1825, Mary moved with her mother to better lodgings and opened a shop called The Fossil Depot, from which she sold numerous of the small fossils she labelled 'curiosities', as well as some parts of the larger creatures, although, of course, her finest finds went to private collectors. For the next twenty years, helped by her mother, who ran the shop while Mary hunted for fossils, Mary continued to amaze learned men with her finds and then be ignored by them when it came to giving credit where credit was due. Her good friend Anna Maria Pinney wrote of Mary's conflicted feelings about this: 'Mary says she stands still, and the world flows by her in a stream, that she likes observing it and discovering the different characters which compose it. But in discovering these characters she takes most violent likes and dislikes ...'

These dislikes were not only to those men who cheated her of recognition, but also to the regular townspeople of Lyme Regis, with whom she had grown up. Many of these might justifiably have thought she was putting on airs with her association with people Pinney called 'those above her'. On the other hand, she had never really been accepted by most in Lyme Regis, who looked down on her unmarried state, her walking about with men of all types on the beaches, and her very strange way of making a living.

Despite her philosophical approach to life as a flowing stream, Mary obviously hungered for recognition. One anecdote is telling. In 1844 Frederick Augustus, King of Saxony, visited Lyme Regis, accompanied by his personal physician, Carl Gustav Carus. Carus had been directed to visit Mary's shop, and he, the king and his retinue were astonished at the fossils that could be found there, everything from 'curiosities' to 'a large slab of blackened clay, in which a perfect ichthyosaurus of at least six feet [1.8 metres] was embedded ... I consider the price demanded—£15—as very moderate'.

Carus then wrote:

> *I was anxious at all events to write down the address and the woman who kept the shop, for it was a woman who had devoted herself to this scientific pursuit, [so she] with a firm hand wrote her name 'Mary Anning' in my pocket-book, and added, as she returned the book into my hands, 'I am well-known throughout the whole of Europe'.*

'The world has used me so unkindly'

Well known or not, Mary was poor, especially as she got older and was unable to move about the cliffs as she had formerly done. She was aided by some welcome charity from a retired British army officer, Lieutenant Colonel Thomas Birch, who was also an amateur fossil collector and who was outraged by Mary's impoverished state—so much so that he sold his entire collection of fossils for the large sum of four hundred pounds, enough to support Mary and her mother, if they were careful, for a good few years. William Buckland managed to get the British Association for the Advancement of Science to begin paying her a small stipend of twenty-five pounds when she was in her late thirties.

Mary's mother died in 1842 and she seems to have become isolated. Her dog, Tray, a beloved animal who clambered over the cliffs with her, was killed in a landslide near Lyme Regis. Several old friends died and Anning seems not to have been close to her brother, Joseph. She brooded on the injustices done to her by many of the famous men of science who had used her findings. When a young girl wrote a fan letter to her from London, she wrote back: 'I beg your pardon for distrusting your friendship. The world has used me so unkindly, I fear it has made me suspicious of everyone.'

In 1845, when Mary was forty-six years old, a lump in her breast turned out to be a malignant tumour. There was no effective treatment at the time. Mary took to palliative measures that apparently included fairly heavy use of laudanum, a combination of opium and alcohol popular as a painkiller. In his profile of her in *All the Year Round*, Dickens wrote: 'Her flying to strong drinks and opium to ease the pain ... her detracting townspeople do not fail to record to her discredit.' But these gossipmongers, Dickens continued, had used Mary as 'bait for tourists' all her life (in fact, as Shelley Emling points out, Mary was singlehandedly responsible for building an entire tourist industry in Lyme Regis at a time when the town was failing).

Mary Anning died on 9 March 1847, possibly alone, and was buried in the nearby Church of St Michael's, near her mother. She was given a stained glass window memorial by the parish 'in commemoration of her usefulness to furthering the science of geology'. Because of her discoveries, scientists had begun to overturn the notion of a world only a few thousand years old and realised that the earth was so many 'ten thousand times ten thousand' years in age, and counting. Charles Darwin's theories of evolution owe a good deal to Mary's humble hammer banging at the blue-grey cliffs of Lyme Regis. She may not have travelled far from her home, but what she discovered travelled a long, long way.

SIR JOSEPH DALTON HOOKER
Adventurous scholar

Joseph Dalton Hooker was one of the most versatile and accomplished botanists who ever lived. During the course of his long life—he was ninety-four when he died—Hooker hunted plants all over the world, from Antarctica to the Himalayas, from the Atlas Mountains of Morocco to the American West, where at the age of sixty he climbed a 4000-metre (13,000-foot) peak in the Rocky Mountains. He also became director of the famous Kew Botanical Gardens outside London, where he oversaw the training of some seven hundred botanists, a veritable horde of flower hunters sent out into the world. Aside from helping make Kew into what one historian has called 'the world's foremost centre for botanical study', he was also a close friend of Charles Darwin; Hooker's careful tabulation of the similarities between widely separated species helped Darwin formulate his theory of evolution. Indeed, it is Hooker's special combination of an adventuring spirit and an astute scientific mind that makes him such an important figure in the history of the natural sciences.

To Antarctica

Joseph Hooker was born in Suffolk, England, in June 1817, the son of Sir William Hooker, the noted Scottish naturalist who acted as a mentor to David Douglas (see page 168) and would become director of the Botanical Gardens at Kew in 1841. Joseph grew up mainly in Glasgow, where he received an early education in the natural sciences while listening to the lectures his father gave to college students. A story often told by Joseph Hooker himself had him wandering away from home at the age of six. When he was found 'grubbing in a wall in the dirty suburbs of the dirty city of Glasgow', he informed his interlocutors that he was searching for the moss *Bryum argentum*. By the age of ten he was so advanced that local schools did not have much to teach him, and he was given advanced tutoring at home.

Like many botanists of the period, he studied for a medical degree, completing his studies at the University of Glasgow and receiving his degree just in time for his first big adventure. The scientist and explorer Sir James Clark Ross, a friend of William Hooker, offered Joseph Hooker a position as physician's assistant aboard one of two ships on a British scientific expedition he was leading to Antarctica to attempt to pinpoint the position of the magnetic South Pole. The twenty-two-year-old Hooker jumped at the chance. Hooker was aboard the *Erebus*; along with the *Terror* it left Great Britain in September of 1839. Their first stop was in the Canary Islands, of which Hooker later enthusiastically wrote: 'There are peculiar emotions on visiting new countries for the first time. I never felt as I did when drawing near Madeira, and probably never shall again. Every knot that the ship approached seemed to call up new subjects of enquiry.'

Unfortunately, Hooker came down with rheumatic fever, but he still managed to collect numerous specimens, so much so that an impressed Ross assigned him a special cabinet at which to work and a table to place his microscope on—not inconsiderable amenities aboard the crowded ships. After recovering, Hooker continued to botanise as the ship made stops at St Helena Island—where Hooker noted local flora being displaced by introduced species—and Cape Town. In March they rounded the Cape of Good Hope into the Indian Ocean, en route to Australia, their staging ground before advancing to the South Pole. On a forlorn island in the Indian Ocean that Captain James Cook had discovered and named Desolation Island more than fifty years before, Hooker re-discovered the anti-scorbutic herb *Pringlea antiscorbutica*, which Cook had used to help prevent scurvy. 'To a crew long confined to salt provisions or indeed to human beings under any circumstances, this is a most important vegetable,' the young Hooker wrote. Cook had said the island had only twenty species of plants, but Hooker found over 150 before the ships sailed on.

A 'small collection of plants'

The expedition arrived in August of 1840 in Van Diemen's Land (Tasmania), where it reprovisioned, and, after a stop in the Auckland Islands, headed south during the short Antarctic summer. Ross led the *Erebus* and *Terror* to latitude 78°3' south, farther than any ship had reached at that time, deep in the Antarctic Circle. In his journal entry for 17 January 1841, Hooker almost ecstatically described the extraordinary Antarctic volcano that Ross named Erebus:

> *The water and the sky were both as blue, or rather more intensely blue than I have ever seen them in the tropics, and all the coast one mass of dazzlingly beautiful peaks of snow, which, when the sun approaches the horizon, reflected the most brilliant tints of golden, yellow and scarlet; and then to see the dark cloud of smoke, tinged with flame, rising from the volcano in a perfect unbroken column; one side jet black, the other giving back the colours of the sun, sometimes turning off at a right angle by some current of wind, and stretching many miles to leeward!*

The ship returned to Van Diemen's Land and then sailed for Sydney, where the crew rested and Hooker received a letter from his father saying that he had been named director of the Kew Gardens;

Joseph Hooker congratulated the elder Hooker by sending him a collection of plants in a Wardian case, one of the first uses of this newly invented terrarium. After this, the *Erebus* and *Terror* set sail again for Antarctica, where Ross finished his scientific experiments. After surviving a storm at sea and a collision between the two vessels—Ross saved the day by steering the wounded *Erebus* between two oncoming icebergs, their huge walls towering over the deck—the expedition headed back to Britain. During a stop in the Hermite Islands, south of Tierra del Fuego, Hooker was fascinated by the number of plant species that seemed quite similar to those back in Britain, which had roughly the same damp, foggy climate as the Hermites.

Hooker had met Charles Darwin briefly before departing on his four-year voyage, and he continued to write to him during the trip, impressed by reading the page proofs of Darwin's book *Voyage of the Beagle*. Hooker arrived back in England in September of 1843, having been away over four years. He had apprised Darwin of his discoveries in the Hermites, for in November Darwin wrote to him that 'I had hoped before this time to have had the pleasure of seeing you & congratulating you on your safe return from your long & glorious voyage'. Through an intermediary, Darwin had sent Hooker a 'small collection of plants' to examine, plants that Darwin had collected in Patagonia and Tierra del Fuego. He then modestly asks Hooker: 'Do make comparative remarks on the species allied to the European species, for the advantage of Botanical Ignoramus'es like myself. It has always struck me as a curious point to find out, whether there are many European genera in T. del Fuego.'

Hooker responded by sending Darwin a detailed analysis, and a close relationship was born—one historian has called Hooker 'Charles Darwin's closest friend and confidant'. Darwin found Hooker's data on the similarities of plant species in different locations extremely helpful in his writing of *The Origin of Species by Means of Natural Selection*, which Hooker read and critiqued at different stages before it was published in 1859.

The Wardian Case

A little-known invention from a little-known man revolutionised plant collecting in the early nineteenth century. That man was Dr Nathaniel Bagshaw Ward, born in London in 1791, the son of a doctor. He was sent to Jamaica at the age of thirteen, and it was probably there that he developed an interest in natural sciences and perhaps also came to understand the difficulties of transporting plant specimens across large expanses of ocean. Plant collectors were never certain that their specimens, no matter how carefully planted in dirt and wrapped in moist burlap, would arrive safely home.

Ward became a doctor, like his father, and practised medicine in East London. He continued to take collecting trips but was dismayed by how the polluted air destroyed his living specimens. One day in 1829, while studying a hawk moth chrysalis, he placed it in a sealed glass jar with leaf mould and some dirt. The chrysalis never sprouted into a moth but, before Dr Ward's curious gaze, something else sprouted: a fern and some grass. Ward decided to see how long they would live. The leaf stayed moist because the water condensed on the side of the glass as the temperature dropped at night, and then dripped back onto the mould (and thus onto the leaf and grass) in the morning.

These plants survived in their sealed environment for four years. Long before that, Ward realised that he had hit upon a solution to the problem of transporting specimens over long distances. He tried an experiment, building large glass cases, framing them in wood and filling them with plants. He sent them on the deck of a ship to Australia, where they arrived after many months in perfect condition. As per his instructions, Australian plants were placed in the cases after they were cleaned out. These arrived back in England in perfect condition. Essentially, Ward had invented the first terrarium, although his cases were known as Wardian cases.

Ward was an acquaintance of William Hooker, who helped spread the news about the cases. Joseph Hooker was among the first to use them, after his trip to Antarctica, and the botanist Robert Fortune employed Wardian cases to send twenty thousand tea plants from Shanghai to India. It was the Wardian cases that made possible the orchid craze of the Victorian era, by transporting these temperamental plants in perfect safety.

'A slender shrub'

Deciding that his main area of interest lay precisely in finding out about the differences and similarities in plant species in widely differing locales, Hooker sought about for another place to which to mount an expedition and found one in Sikkim, the tiny kingdom in the northeast Himalayas, whose 'ground [was] untrodden by traveller or naturalist'. Hooker went on to write that its rajah, or ruler, was 'all but a dependant of the British government, and, it was supposed, would therefore be glad to facilitate my researches'. This would turn out to be a serious misjudgment but, for any number of reasons at the time, Sikkim seemed to be the right place to go. It was unexplored by naturalists; Hooker's friend Hugh Falconer was about to become superintendent of the Calcutta Gardens and could thus be counted on for assistance; and Hooker's father was willing to commission him to collect plants for Kew.

Hooker, unlike less fortunate botanists, lived in a world of felicitous connection. Leaving Southampton for the voyage to India in September of 1847, he met Lord Dalhousie, the newly named governor-general of India. Once they got to Suez, Dalhousie, as Hooker wrote, 'did me the honour of desiring me to consider myself in the position of one of his suite for the remainder of the voyage', thus allowing Hooker to travel in style on a steam frigate sent from India to pick up Dalhousie. Once in Calcutta, Dalhousie, apparently much taken with Hooker, allowed him to stay at his official residence, as Hooker noted in gratitude, 'giving me such a position near himself as ensured me the best reception everywhere; no other introduction being needed'.

From Calcutta Hooker travelled to the Himalayan foothills, arriving in April of 1848 in Darjeeling, where he was immediately befriended by Brian Hodgson, a noted expert on Indian zoology, and Dr Archibald Campbell, the British political agent to Sikkim, who now began to try to secure permission for Hooker and his expedition to enter Sikkim. It turned out to be a torturous process. The rajah of

Sikkim had been helped by the British thirty years earlier, when they had put him back on his throne after the Nepalese had tried to oust him, but he was now ageing and the real power behind his throne was the dewan, or prime minister, a Tibetan who was a relative of one of the rajah's numerous wives and, as Hooker wrote, 'unsurpassed for insolence and avarice'. The dewan disliked and distrusted the British and counselled the rajah to delay the expedition.

While the negotiations were continuing, Hooker had to content himself with taking plant hunting expeditions into the Himalayan foothills, where he had the consolation prize of finding some glorious rhododendrons, including a beautiful one he named *Rhododendron dalhousiae*, after Lady Dalhousie, Lord Dalhousie's wife—'a slender shrub, bearing from three to six white lemon-scented bells, four and a half inches [115 millimetres] long and as many broad, at the end of each branch'.

'To travel to the Snowy passes'

With negotiations still stalled in the autumn of 1848, Campbell sought the help of Lord Dalhousie, who wrote to the rajah to ask 'full leave [for Hooker] to travel to the Snowy passes'. The rajah, prodded by the dewan, flatly refused, at which point Dalhousie threatened to invade Sikkim and the rajah was forced to relent—it was now less about Hooker going to Sikkim than about what the British obviously considered to be insubordination on the part of the locals. An agreement was worked out that would send Hooker first to Nepal to explore the northeast part of that mountain kingdom, from which point he would be allowed to return by way of Sikkim.

On 27 October 1848, he set out. His expedition numbered fifty-six in all. He was accompanied by guards—from Nepal, as the rajah there, sensing an opening, was quite helpful—cooks and porters, as well as Indian naturalists (both bird and plant collectors) Hooker had trained. Hooker, who carried with him the prejudices of his time, thought that his Lepcha porters were amiable, if sometimes lazy, but the Bhutanese who also worked for him he considered 'worthless'. He climbed ever higher in the mountains, crossing seemingly bottomless gorges and rushing mountain streams by means of terrifying suspension bridges: 'In these bridges,' he wrote, 'the principal chains are clamped to rocks on either shore, and the suspended loops occur at intervals of eight to ten feet [2.5 to 3 metres]; the single sal-plank laid on these loops swings terrifically, and, the handrails, not being four feet [1.2 metres] high, the sense of insecurity is very great'.

Hooker reached the village of Wallanchoon in late November. The inhabitants there were 'good-natured, intolerably dirty Tibetans', according to Hooker. They raised hardy yaks and owned large and ferocious mastiff dogs. They were reluctant to escort Hooker up to the Wallanchoon Pass (at 4900 metres, or 16,000 feet) so that he could take certain measurements before the snow became too heavy, but finally agreed after Hooker applied a combination of 'bullying ... persuasions, and the prescribing of pills, prayers and charms ... for the sick of the village, whereby I gained some favour'.

Climbing to the top of the pass, Hooker found 'huge angular and detached masses of rock ... scattered about, and to the right and left snowy peaks towered over the surrounding mountains'. Despite suffering altitude sickness ('lassitude, headache and giddiness'), Hooker did discover, to his 'deepest interest', two of the commonest of all British weeds, a grass (*Poa annua*) and shepherd's purse (*Capsella bursa-pastoris*), which he thought could not be native there but had probably been dropped inadvertently by travellers. Hooker went on to reflect, while writing his two-volume *Himalayan Journals, or Notes of a Naturalist*:

> *At this moment, these common weeds more vividly recall to me that wild scene than does my journal, and remind me how I went on my way, taxing my memory for all it ever knew of the geographical distribution of the shepherd's purse, and musing on the probability of the plant having found its way thither over all Central Asia ...*

A favoured Fortune

Most plant hunters of the Victorian era took enormous risks while searching for their beloved specimens, but Robert Fortune was one of the few who seemed to seek out risks with relish. Fortune was born in Scotland in 1812. Not a good deal is known about his early life, but by 1840 he had begun working at the Botanic Garden in Edinburgh and was eventually hired by the British Horticultural Society as one of their collectors, which means he was sent to exotic places in search of exotic plants.

His first assignment was China, after the Treaty of Nanking ended the Second Opium War in 1842. Fortune was sent out to seek blue flowering peonies and to see what kind of peaches grew in the emperor's garden. Foreigners, especially the British, were not exactly welcomed in China, and Fortune was paid the miserable sum of one hundred pounds to risk his life for fruit and flowers. When he complained, an official at the society told him: 'The mere pecuniary returns of your mission ought to be but a secondary consideration to you.'

Undaunted, Fortune arrived in Hong Kong and spent three years collecting plant specimens. It is a wonder he survived this journey at all. At one point, on the outskirts of Canton, he was surrounded by an angry mob that may have been trying to rob him; he barely escaped being stoned to death. Later, travelling by junk to Shanghai, and ill with a fever, he fought off several pirate attacks by waiting until the marauders were about to board the boat and then cutting them down with blasts from a double-barrelled fowling piece.

On an 1848 trip for the East India Company he managed to send back to India thousands of Chinese tea plants (in Wardian cases) and helped establish the tea industry in India. Once again, he singlehandedly fought off mobs and pirates. 'His courage was phenomenal,' writes Alice M. Coats in her book The Plant Hunters. *'But it sprang partly from his conviction that the Chinese could not be his superiors in anything but numbers.'*

Fortune was smart as well as brave. At the end of his career, he made a great deal of money selling Chinese and Japanese objets d'art *that he had collected during his numerous travels and was thus able to fund nineteen years of comfortable retirement.*

'Where no European had ever been'

After returning to Wallanchoon, Hooker sent all but nineteen of his men back to Darjeeling, keeping only those souls hearty enough for 'a journey of great hardship' deep into the mountain passes through which the Yangma River ran. 'The scenery was wild and very grand, our path lying through a narrow gorge, choked with pine trees, down which the river roared in furious torrent ... The road was very bad, often up ladders, and along planks lashed to the faces of precipices ...'

Despite the ruggedness of this journey, Hooker found numerous specimens of hardy rhododendrons and larches in the Yangma Valley and near Mount Nango. He attempted to cross the Kanglanama Pass, but found it clogged with snow, which, as late as the season was, had started to become a problem—Hooker had begun propping a surveyor's tripod over his head in his tent, to leave himself breathing space in case of an avalanche. He now headed southwest towards the Khabil Valley, attempting to find entry into Sikkim through another mountain pass. But by now he had started having problems finding food in the villages through which he passed. He found out, to his dismay, that the trusted Nepalese officer in charge of the Gurkhas who formed his bodyguard had been stealing all the food given to the group as tribute—the Gurkhas, therefore, had a good deal to eat, while Hooker and his collectors were starving.

Fortunately, Hooker was able to cross the Islumbo Pass into Sikkim, arriving in the village of Lingcham on 7 December, pleased to find 'an abundance of milk, eggs, fowls, plantains, and Murwa beer'. He also found welcome news: Dr Archibald Campbell, the British political officer, had at last been allowed to enter Sikkim to talk with the rajah and would like Hooker to join him at Bhomsong on the Tista River—'where no European had ever been'. Hooker set out happily to meet his friend on 20 December, arriving at Bhomsong a few days later and warmly embracing Campbell. All seemed to be going well. The rajah sent Hooker a present of 'eighty pounds

[36 kilograms] of rancid yak butter, done up in yak-hair cloth', as well as numerous other foodstuffs. That evening they met the dewan, described by Hooker as 'a good-looking Tibetan [with] a very broad Tartar face, quite free of hair [and] a small and beautifully formed mouth'.

The dewan's manners were 'courteous and polite', Hooker wrote, but he felt that they were a mask for the dewan's true feelings, which were hatred of the British. Wanting to control trade with Sikkim, the dewan, according to Hooker, was a 'mere plunderer ... trading in great and small wares' that the British could provide far more cheaply. Hooker stayed with Campbell until 2 January 1849 and then turned back through Sikkim to Darjeeling, arriving on 19 January with '80 coolie-loads' of specimens, some of them geological specimens but many rhododendron seeds that he had gathered at altitude 'and with cold fingers it is not easy at the ripening season, December, to collect those from the scattered twigs, generally out of reach'.

Wild and magnificent scenery

As soon as possible after unloading, categorising and sending his specimens to his father at Kew, Hooker set off again for Sikkim. In May of 1849 he left Darjeeling, following the Tista River to its headwaters and into Sikkim but finding his way considerably impeded by the forces of the dewan. Although the rajah had given permission for this second trip, the dewan now strove mightily to make sure it was not a success. He forbade villagers to sell Hooker and his party any food, something Hooker had thought might be a problem—so the botanist had arranged for shipments of rice to be sent to the expedition from Darjeeling. However, the dewan was one step ahead of him—he had expressly forbidden any repairs to be made on roads and also ordered that his men destroy bridges, remove pavement stones and fell trees over paths.

This made travel for Hooker and his party extraordinarily dangerous, especially since it was the rainy season. Because of the dewan, it took Hooker eighty-three days (instead of thirty) to travel through Sikkim to the Kongra Lama pass into Tibet, which was one of his destinations. For eight days straight, all he had to eat was rice and chilli vinegar. Hooker sent repeated letters to Campbell (who was still in Sikkim) complaining of his treatment. Campbell would take them to the rajah, who would sympathise, send Hooker useless presents (such as silk robes) and order that he be fed and taken care of. Then the dewan would countermand the orders and the whole cycle would begin again.

But, making his way through wild and magnificent scenery, Hooker finally entered Tibet, spending a week at Tungu, just below the summit of the Kongra Lama pass, where he was welcomed quite cordially—he was the first European to have been there—by men 'lolling' their tongues and pulling at their ears, which struck Hooker as highly unusual but turned out to be a traditional Tibetan greeting. Hooker continued botanising, finding ten types of rhododendron, including two new species. He also found a highly toxic type of rhododendron, *Rhododendron cinnabarinum*: when eaten by the goats and kids accompanying the expedition, the animals 'died foaming at the mouth and grinding their teeth'.

Finally returning into Sikkim, the intrepid Hooker climbed the 5880-metre (19,300-foot) Mount Donkia, becoming—albeit briefly—the world's highest climber (unbeknownst to him). Standing on the mountain, he had the great good fortune to experience an unusual atmospheric phenomenon called the 'Brocken spectre', in which his enormous, magnified shadow was cast on the undersides of the clouds by the setting sun, his head crowned by golden rays, making it appear as if he were a giant astride the world, wearing a glowing halo.

It was a moment he never forgot and one he hung onto during the events that followed. On 2 October he was reunited with Archibald Campbell, who had been in Sikkim attempting to convince the rajah of the dewan's treachery. Campbell was convinced by an enthusiastic

Hooker to follow him back into Tibet via the Kongra Lama pass and then re-enter Sikkim by the Donkia pass, which they did, gathering the seeds of twenty-four species of rhododendron along the way. Back in Sikkim, they headed together to the capital town of Tumlong, where they tried to personally interview the rajah, but they were turned away by the dewan's officials.

Hooker and Campbell left, journeying back towards Tibet. On 7 November, as they camped for the night, Hooker heard Campbell shout: 'Hooker! Hooker! The savages are murdering me!' Racing to his friend's tent, Hooker saw numerous men, who turned out to be in the employ of the dewan, beating Campbell, whose arms had been tied behind his back.

Kew Gardens

Kew Gardens, which Sir William Hooker and then his son, Joseph Dalton Hooker, so ably directed for so many years, was originally a 4.5-hectare (11-acre) patch of 'exotic garden' within the larger Richmond Gardens, where members of the royal family maintained a residence in what is now southwest London. King George III improved, enlarged and re-landscaped the gardens, while Queen Charlotte, his wife, sought plants from such naturalists as Sir Joseph Banks. Banks became, unofficially, the first director of the royal garden at Kew, consulted by the king and queen on almost all aspects of the preserve's care and maintenance. In the early nineteenth century George and Charlotte added six greenhouses, where exotic specimens sent by collectors from the far corners of the burgeoning empire thrived.

Unfortunately, after the death of Joseph Banks, Kew Gardens fell into a state of disuse and disrepair; when William Hooker took over as director in 1841, the glasshouses were about to be turned into vineries. It was under his auspices, and later those of his son, that the Royal Botanical Gardens at Kew underwent its reawakening and revival, with plants sent by plant hunters from all over the world. Today the gardens cover more than 120 hectares (300 acres) and receive over two million visitors a year.

The great adventure

Hooker was kept from going to Campbell's aid but otherwise was not harmed. He was informed that Campbell was now a political prisoner of the rajah, something Hooker did not believe for a moment—he knew the dewan was behind the assault. He was told he was free to go, but he refused, instead following the dewan's men as they led their prisoner back to Tumlong. There, Hooker was informed that Campbell was to be held as a hostage until the British rulers of India made concessions to the dewan, concessions that included an agreement not to trade in Sikkim and the relaxation of the rules against slavery, which had been outlawed in the British Empire.

Rather ironically, the British Raj was unaware that Campbell was a captive because the only letter sent on the matter—Hooker and Campbell were forbidden to write—was written in Tibetan. When it arrived at Darjeeling, it was read to Campbell's secretary by a translator, who, unfortunately, omitted the last line, which related that Campbell was a hostage. Assuming it was just another list of Sikkimese demands, the secretary placed it on Campbell's desk for him to deal with on his return. When the Sikkimese became nervous after the British did not reply, Hooker was allowed to write, but still more time went by because he addressed his letter to Lord Dalhousie, who was travelling.

When the British finally realised what was happening, they reacted as anyone (but apparently not the dewan) could have predicted: they sent a letter demanding Campbell's release. And they sent a military force to Darjeeling, ready to invade Sikkim. The alarmed rajah finally understood the dewan's perfidy, fired him and ordered that Campbell and Hooker be released. The dewan was forced to take them to Darjeeling himself. While still a few kilometres outside the city, the two Britons realised that the dewan was terrified of even approaching the British forces and feared that he might kill them out of panic. They convinced him to give them horses and let them ride the last few kilometres into Darjeeling on their own. It was Christmas Eve, 1849.

The dewan's plan had been an abject failure. As further punishment, the British annexed the lower part of Sikkim, the only really fertile part of the country (perfect for growing tea), making it part of their Indian Empire. Hooker lost all his instruments and certain of his specimens—which the British government somewhat churlishly refused to reimburse him for—but in the main was content. He had been on the great adventure of his life.

A hero's welcome

Joseph Hooker finally arrived home in March of 1851 after three and a half years in India—he was given a hero's welcome for his role in sticking by his friend Campbell. He then married his fiancée, Frances Harriet Henslow, and wrote books about his journey to Antarctica and his Himalayan adventures—*Himalayan Journals* became an immensely popular title. Despite the fact that he was to have nine children by two wives, Hooker managed to find time to make further botanising trips to Morocco and the United States, to publicly defend Charles Darwin's *On the Origin of Species* when the book was published and, most importantly, to take over the direction of the Botanical Gardens at Kew, where he managed to increase the number of visitors from nine thousand annually in 1841 to three million by the turn of the century. He was knighted in 1877 for his contributions to Great Britain. And, as a last, great masterwork, Hooker, writing with botanist George Bentham, completed *Genera Plantarum*. Twenty-six years in the writing, 3363 pages long (with over 1,600,000 words), the book identified every known plant in the world and became known to generations of botanists simply as 'Bentham & Hooker'.

When he died in his sleep at the age of ninety-four in 1911, Sir Joseph Dalton Hooker had ensured he would be remembered as probably the most important botanist of the nineteenth century.

HENRY WALTER BATES
'a solitary stranger ... on a strange errand'

Not every natural historian of the nineteenth century had the publication of his journals greeted by 'An Appreciation' written by Charles Darwin, but in 1863 that is exactly what happened to Henry Walter Bates. Chronicling Bates's adventures in the Brazilian jungle, as recounted in *The Naturalist on the River Amazons*, Darwin writes of Bates's 'watchful intelligence' and exclaims that he is 'obviously a man of no ordinary mark'. With some wonder, Darwin counts the number of specimens Bates brought back to England after eleven years in South America: 'It has been ascertained that representatives of no less than 14,712 species are amongst them, of which about 8,000 were previously unknown to science.'

But even beyond this: 'Mr. Bates does not confine himself to his entomological discoveries, nor to any other branch of Natural History, but supplies a general outline of his adventures during his journeyings up and down the mighty [Amazon], and a variety of information concerning every object of interest, whether cultural or political, that he met with by the way.'

In other words, Bates was not merely some scientific drudge, his nose buried in Amazonian bushes recording the love life of ants (although there was definitely that side to him, as well) but a recorder of the broad sweep of human life as it collided with nature in that most extraordinary of the world's mixing bowls: the Amazon river basin.

'Mobbed by Curl-Crested Toucans'

Henry Walter Bates was walking through the forest in the Ega district one day when he spied, on a high branch above him, a curl-crested toucan, a quite common and colourful Amazonian bird that was hunted locally for its delicious flesh. Deciding on the spot to shoot it down for his supper, he raised his rifle, aimed, and hit the bird, which came crashing down through the tree branches into the underbrush.

But the bird was only wounded. When Bates went to pick it up from its landing spot, it 'set up a loud scream'. To his horror, Bates suddenly realised that 'the shady nook' surrounding him was alive with curl-crested toucans, all of which had begun to move menacingly, in what could be seen as a show of allegiance to the injured toucan, in his direction. 'They descended towards me, hopping from bough to bough, some of them swinging on the loops and cables of woody lianas, and all croaking and fluttering their wings like so many furies.If I had a long stick in my hand, I could have knocked several of them over.'

In a scene prefiguring the Alfred Hitchcock movie The Birds, *Bates frantically loaded his gun and prepared to do battle with the vengeful creatures. But then the bird he had wounded died and, 'the screaming of their companion having ceased, [the toucans] remounted the trees, and before I could reload every one of them had disappeared'.*

The illustration of this incident, captioned 'Mobbed by Curl-Crested Toucans', was the frontispiece of The Naturalist on the River Amazons *and wonderfully captures the incident. The bespectacled Bates, dying toucan in hand, stares up in horror as the birds crowd closely around him, large beaks snapping.*

'On coleopterous insects frequenting damp places'

The frontispiece illustration of *The Naturalist on the River Amazons* became famous after the book's publication as a sort of generic depiction of nineteenth century British travellers in the wild—indeed, it is sometimes reproduced today without reference to the fact that it is from Bates's book and, in fact, illustrates a famous incident concerning the author and a flock of threatening birds. But it captures Bates to a tee—the thick glasses, floppy hat, bushy moustache, long-sleeved checked shirt and baggy trousers. Indeed, super-geek, in his natural environment.

It is actually a wonder that this bespectacled, sickly (Bates was prone to recurring bouts of bronchitis, the disease that would eventually rob him of his life) son of a Leicester hosiery manufacturer ever made it out of England at all. He was born in 1825 to parents who struggled financially, although they raised their children with a strong emphasis on reading and writing. Bates was forced to leave school at the age of thirteen, to be apprenticed to a wholesale merchant who was a colleague of his father and for whom he worked thirteen hours a day, six days a week. Even so, Bates's interest in nature, particularly in insects, which had begun when he was a very young child, continued unabated, as he spent his spare time searching nearby Charnwood Forest for specimens of beetles and ants. As he grew older, he began to take courses at the Mechanics' Institute (an educational organisation that offered learning to those of lower incomes). When he was eighteen, he began contributing articles to a well-known scholarly journal, *The Zoologist*—his first publication was a paper on beetles entitled 'On coleopterous insects frequenting damp places'.

Still, he might have remained merely another fervid amateur naturalist—a type abounding in Britain in those days—had it not been for a fortuitous meeting in 1844 in the Leicester public library with a young teacher at the Collegiate School in Leicester, Alfred Russel Wallace. Two years older than Bates and destined to become one of the great natural scientists of nineteenth century England,

Wallace was tall, as socially awkward as Bates and equally fascinated with the natural world. Wallace, too, was from a poor family and had been educated mostly at the Mechanics' Institute. One day at the library, he and Bates fell into conversation and Bates, seeing that they had similar interests, offered to show Wallace his collection of beetle specimens. As Wallace was to write in his 1905 memoir, *My Life*, he was astonished at 'the great number of beetles, their many strange forms and other beautiful markings or colouring, and was even more surprised when I found out that almost all I saw had been collected round Leicester'.

'The great primaeval forest'

Bates and Wallace soon became fast friends and went on as many natural excursions through the English countryside as time would allow. They had both read the work of Alexander von Humboldt (see page 95) and longed to travel through the wilds of South America. They were also fascinated by Charles Darwin's depiction of his circumnavigation of the world in *Voyage of the Beagle* (published in 1839) and both men speculated about 'solving the problem of the origin of species'.

By 1847 Bates was working as an office clerk in a town near Leicester, and feeling stifled and unfulfilled. He and Wallace read a book entitled *Voyage up the River Amazon, Including a Residence at Para*, by the American lepidopterist W. H. Edwards, and were fascinated by Edwards's depictions of 'the beauty and the grandeur of the vegetation' and the 'kindness and hospitality of the people' of Pará, a town 110 kilometres (70 miles) from the mouth of the Amazon River. Both men took time off from their humdrum jobs to call upon some of the leading scientific names of their day and ask their advice. These men (who included Edward Doubleday, curator of lepidoptera at the British Museum, and Sir William Hooker, head of the Botanical Gardens at Kew) advised them that private collectors would pay a great deal for specimen collections from places like the

Amazon. With astonishing speed, Wallace and Bates secured several commissions that would easily pay for their trip. The two young men quit their jobs and set off from Liverpool on 26 April 1848 aboard a small trading vessel. Thirty-two days later, they arrived off the coast of Brazil, anchoring offshore 'at the mouth of the vast region watered by the Amazon'.

Awestruck, Bates borrowed the captain's spyglass. 'To the eastward,' he later wrote, 'the country was not remarkable in appearance, being slightly undulating, with bare sandhills and scattered trees; but to the westward, stretching towards the mouth of the river, we could see ... a long line of forest, rising apparently out of the water, a densely-packed mass of tall trees ... This was the frontier ... the great primaeval forest characteristic of this region, which contains so many wonders in its recesses, and clothes the whole surface of the country for two thousand miles [3200 kilometres] from this point to the foot of the Andes.'

'Natural riches and human poverty'

Bates and Wallace sailed up the Para River—at some places 58 kilometres (36 miles) wide—and arrived at the town of Pará (present-day Belem) on 28 May. Pará was an old city, founded in 1615 by the Portuguese, that had once been a bustling mercantile centre but that had fallen into disuse and disrepair over the years. 'The wooden palings which surrounded the weed-grown gardens were strewn about,' Bates wrote, 'and hogs, goats and ill-fed poultry wandered in and out through the gaps.' Yet Bates—a man with an appreciative eye for female beauty—also described 'several handsome women, dressed in a slovenly manner, barefoot or shod in loose slippers; but wearing richly decorated ear-rings, and around their necks strings of very large cold beads. They had dark expressive eyes and remarkably rich heads of hair.'

It seemed to Bates that the contrast these women provided— beauty with squalor—was symbolic of the 'mixture of natural riches and human poverty' that characterised the area. Finding a house and establishing themselves with the small European community in Pará, Bates and Wallace soon found the town to be an attractive place— much more so than England, whose dank climate and hard work had so stifled the young men. Even the gecko lizards on the walls were fascinating to Bates—although 'very repulsive in appearance', their feet were 'beautifully adapted for clinging to and running over smooth surfaces'.

But the real bounty came when Bates and Wallace ventured out of Pará. Within an hour of the town, they found an astonishing seven hundred species of butterfly, 'whilst the total number found in the British Islands does not exceed sixty-six'. Eighteen species of rare (in the British Isles) swallowtail butterfly were found within ten minutes walk of their house. The two men soon settled into a routine. Their house on the outskirts of Pará was a square, cool building with a tiled roof and a large veranda. There was forest on three sides of them. Bates and Wallace would arise at dawn, drink coffee prepared by their Brazilian servant, Isidore, and then head out into the woods for two hours of birding before breakfast. It was an idyllic scene: 'The heavy dew or the previous night's rain, which lay on the moist foliage, becoming quickly dissipated by the glowing sun ... All nature was fresh, new leaf and flower-buds expanding rapidly.'

After hunting for birds and stopping for breakfast, Bates and Wallace spent from 10 am to 3 pm studying 'the insects of the forest' before the most powerful heat of the day set in and they returned home to rest. In the summer, storm clouds would appear on the horizon: 'The heat and electric tension of the atmosphere would then become almost insupportable. Languor and uneasiness would seize on every one [until] the whole eastern horizon would become almost suddenly black' and rain would come crashing down with a mighty wind and flashes of lightning. And the next morning the entire cycle would repeat itself.

Bates and Wallace spent eighteen months around Pará, becoming fast friends and collecting numerous specimens. When they tired of their wanderings, they would head into town 'to see Brazilian life or enjoy the pleasure of European and American society'.

Alfred Russel Wallace

Bates's companion for the first part of his Amazon stay, Alfred Russel Wallace, would easily surpass him in fame. In 1850 Bates and Wallace parted company and Wallace ascended the Negro River, collecting fabulous specimens. In 1852, weakened by malaria, he was ready to return to England. He left Brazil on 12 July on the brig Helen, but after twenty-eight days at sea the ship's cargo of dry balsa wood caught fire, and the Helen sank. Wallace's valuable collection of specimens was lost and he spent ten days at sea in an open boat before being picked up.

Undaunted, Wallace published two books about his Amazonian adventures and then headed to the Malay Archipelago, where he stayed until 1864. There he began to do increasingly important work, collecting more than 125,000 specimens. He noted marked differences between animals living on the western islands of the archipelago and those on the eastern islands—one group of animals of Australian origin, the other Asian—and the dividing line between the groups is now known as the Wallace Line. Increasingly, he began formulating a theory of natural selection, the same theory that his and Bates's hero Darwin had been working on for years. In 1858 Wallace sent Darwin an outline of his theories on the origin of species, which prompted Darwin to arrange to publish what he called Wallace's 'short abstract' along with unpublished work of his own, in order both to give credit to Wallace and to show that Darwin had been working ahead of him on the same idea (the publication of On the Origin of Species *was still a year away). While there were differences in their ideas on natural selection—Wallace emphasised the role of environment in species change, while Darwin focused on competition—Wallace, large-minded and generous, thereafter became a strong supporter of Darwin and his theories.*

Wallace died in 1913, covered with honours, at the age of ninety.

Serpents and ants

Bates's book recounting his experiences, *The Naturalist on the River Amazons* (Bates used a literal translation of the Portuguese name for the river, *Rio Amazonas*, hence the plural in the title), is filled with casual descriptions of somewhat hair-raising episodes. Boa constrictors were quite common around Pará, so much so that Bates simply got used to them and was not surprised when a local lamplighter showed him one he had cut in half a few metres from Bates's veranda. Once when he was out in the jungle, Bates stepped on the tail of a young and very poisonous serpent, the *jararaca*, which immediately 'turned round and bit me on the trousers'. Fortunately, a young Indian travelling with Bates killed the snake before it could disentangle itself and attack in earnest. There were so many snakes in the woods that Bates would turn a corner and find one at eye level, curled around a branch. 'It was rather alarming in retrospect in entomologizing about the trunks of trees, to suddenly encounter on turning round, as sometimes happened, a pair of glittering eyes and a forked tongue within a few inches of one's head.' His descriptions of the creatures are often poetic. Once, he startled a boa constrictor. 'On seeing me the reptile suddenly turned, and glided at an accelerated rate down the path,' he wrote. 'The rapidly moving and shining body looked like a stream of brown liquid flowing over the thick bed of fallen leaves.'

Another fascinating inhabitant of the forest was the ant, of which there were multiple varieties. On their initial walks around Pará, Bates and Wallace were surprised to see long processions of what appeared to be 'animated leaves' walking down jungle paths. These were actually long lines of Sauba ants, whose chief goal in life was climbing trees and bushes, tearing off leaves and carrying them to their large, mounded anthills. Unfortunately, as Bates points out, 'this vast host of busy diminutive labourers, favoured not wild trees, but cultivated ones, for their purpose, especially coffee and orange trees'. The ants were also in the habit of robbing granaries and storerooms. Once, Bates was awakened by a servant who told him that rats were at work plundering his store of *farinha* grain, but when he entered his supply room he found, instead, 'a broad column of Sauba ants, consisting of thousands of individuals, as busy as possible, passing to and fro between the door and my precious baskets'. Bates and his servant tried to stomp out the creatures with the wooden clogs they wore around the house, but more ants simply arrived to take the place of those that died. The creatures returned in force the next night and Bates was forced to resort to the expedient of laying 'trains of gunpowder' along their line of march, lighting them, and blowing the ants to smithereens. Even this drastic step had to be repeated numerous times before the creatures were willing to leave Bates's grain alone.

No one at that time knew what the Sauba wanted with the leaves they harvested so assiduously. Bates observed them dragging the leaves back to their homes and assumed the ants wanted the vegetation to 'thatch the roofs of their subterranean dwellings', but in fact scientists have since concluded that the Sauba are actually farming. Left within the nest and covered with earth, the leaves degenerate into fertiliser, which is used to grow a fungus eaten by the ants as their main source of food.

'This strange solitude'
In 1849 Bates and Wallace made a journey up to Tocantins River, which feeds into the Amazon and, at 2500 kilometres (1600 miles) long, is the third largest river in the Amazon tributary system. It was on this journey that Bates found that what he thought was an Amazon legend—the bird-eating spider—was, in fact, a reality. He discovered a 'large hairy spider of the genus Mygale', whose legs were as long as 18 centimetres (7 inches) and which was covered with disgusting red and grey hairs that, when touched, 'caused a peculiar and almost maddening irritation' on human skin. Near the spider, tangled in the creature's web, were two finches, one dead, the other almost so, which were 'smeared with the filthy liquor or saliva exuded by the monster.

I drove away the spider and took the birds, but the second one soon died.' Bates also related that some mygales were huge: 'One day I saw the children belonging to an Indian family, who collected for me, with one of these monsters secured by a cord round its waist, by which they were leading it about the house as if it were a dog.'

When Bates and Wallace returned to Pará in late 1849, they prepared a large number of specimens and sent them off to England, and then, in 1850, the two men parted company, finding it, as Bates wrote, 'more convenient to explore separate districts and collect independently'. Wallace went to journey up the Negro River, the Amazon's largest tributary, while Bates ascended the Amazon, going deep into the interior near the village of Ega (present-day Teffe), where he spent a year collecting before becoming sick with yellow fever and returning to Pará. Out of funds, he thought he would have to return to England, but, unexpectedly, a large sum arrived, his share of the profits from the sale of the collection of specimens he and Wallace had shipped home. Regaining his health to some extent, Bates then decided to go back up the Amazon, this time to the town of Santerém. After spending nearly four days sailing up the river in 'an ill-rigged' sailboat, he came, in November of 1851, to 'a pretty little town' of 2500 inhabitants, situated at the mouth of the Tapajos River. It was a place of 'broad white sandy beaches, limpid dark green waters, and [a] line of picturesque hills rising behind over the fringe of the green forest', and it would be Bates's base for the next several years.

From Santerém, Bates travelled deep into the Amazonian forest in all directions. Here it can be said that his true solitary foraging for species began. Bates at this time often collected alone and he could feel the immense loneliness of nature: 'The lazy flapping flight of large blue and black morpho butterflies high in the air, the hum of insects, and many inanimate sounds, contributed their share to the total impression this strange solitude produced.' At the same time, he was surrounded by life—by birds, insects, toads, monkeys (the 'hollow cavernous roar' of the howler monkey echoing wherever

Bates walked in the forest), as well as the omnipresent snakes and seldom-seen jaguars. And just when it was all merging into one grand cacophonous roar, he discovered some individual creature that astonished him—journeying up the Tapajos, he found a unique cricket whose chirp 'was the loudest and most extraordinary I ever heard produced by an orthopterous insect'.

The Upper Amazon

After three and a half years at Santerém, Bates now decided to head far into the interior, to the Upper Amazon, to revisit Ega and explore the country further. Carried in a large canoe, through a place where the river had 'an evil reputation for storms and mosquitoes', Bates now entered a country of plains covered with endless jungle that was far hotter and more humid than that around Santerém, because the trade winds did not reach this far west. 'One lives here as in a permanent vapour bath', Bates wrote. While still on the river, Bates was assailed by the pium fly, which attacked all the exposed parts of his body for the first week—'by that time', Bates later wrote, he was 'so closely covered with black punctures that the little bloodsuckers could not very easily find an unoccupied place to work on'. Mosquitoes were also an intense nuisance, leaving small infected welts and cuts from their bites.

Despite this, Bates felt at home when he settled in Ega. He was not easily scared off. In one typical passage in his journals, he describes the difficulty of bathing in the river near Ega.

Alligators were rather troublesome in the dry season. During these months there was almost always one or two lying in wait near the bathing place for anything that might turn up at the edge of the water; dog, sheep, pig, child or drunken Indian. I used to imitate the natives in not advancing far from the bank and in keeping my eye fixed on the monster, which stared with a disgusting leer along the surface of the water ... I was never threatened myself, but I often saw the crowds

of women and children scared while bathing, by the beast making a
movement towards them; a general scamper to the shore and peals of
laughter were always the result in these cases.

In the next four years in the jungle around Ega—and also on excursions that took him as far as the Peruvian border—Bates collected fully seven thousand of the fourteen thousand species he would catalogue during his time in Brazil, including 550 species of butterfly alone. It was along the Upper Amazon that he 'had the great pleasure of seeing the rare and curious Umbrella Bird', a species of bird about the size of a crow, with clumps of shaggy hairs atop its skull. Within the hairs were long bony quills 'which, when raised, spread themselves out in the form of a fringed sun-shade over the head'. It was around Ega, too, that Bates had his famous adventure with toucans and it was also here that he found himself assailed one night by vampire bats. 'I seized them, as they were crawling all over me, and dashed them against the wall,' he wrote. He felt that these creatures were essentially harmless but was not pleased to discover, the morning after the bat attack, a wound on his thigh: 'This was rather unpleasant', he wrote, with typical understatement.

Finally, it was on the Upper Amazon that Bates began to understand some of the fearful Indian myths about the jungle. Bates was eminently and always a man of science, logical and calm, and yet—there were sounds coming out of the Amazon that were, as he wrote, 'impossible to account for'. They included horrible screams and noises that sounded 'like the clang of an iron bar against a hard, hollow tree'. The Indians attributed this to Curupira, the Amazonian Bigfoot or Yeti, a shaggy creature with backward pointing feet to ward off trackers. Bates also learned of a huge and horrible serpent called the Mai d'Agua, which frequented the river and gobbled unwary canoeists whole. While he never ran into this Loch Ness-like monster, Bates did have the unsettling experience of having an enormous anaconda rise out of the river one calm night, destroy the poultry cage at the back of his canoe and swallow two chickens before disappearing back into the dark water.

'Nature alone is not sufficient'

By early 1859, after eleven years in the Amazonian jungle, Bates had finally had enough. Wracked with malaria, he boarded a ship at Ega with all his collections and travelled back to Pará, having been away over seven and a half years. He was conflicted about the idea of returning to England. On the one hand, he wrote, after numerous hardships and much time spent in solitude, 'I was obliged, at last, to come to the conclusion that the contemplation of Nature alone is not sufficient to fill the human heart and mind.' Often, alone in the jungle, he had considered himself 'a solitary stranger alone on a strange errand'.

But, still, the thought of returning to civilisation frightened and disturbed him. At the end of *The Naturalist on the River Amazons*, he penned a brilliant, elegiac passage that has become famous. It was his last night at Pará, and his thoughts turned to home.

> *Recollections of English climate, scenery, and modes of life came to me with a vividness I had never before experienced during the eleven years of my absence. Pictures of startling clearness rose up of the gloomy winters, the long grey twilights, murky atmosphere, elongated shadows, chilly springs, and sloppy summers; of factory chimneys and crowds of grimy operatives, run to work in early morning by factory bells; of union workhouses, confined rooms, artificial cares, and slavish conventionalities. To live again amidst these dull scenes I was quitting a country of perpetual summer, where my life had been spent like that of three-fourths of the people in gypsy fashion, on the endless streams or in boundless forests.*

Nonetheless, Bates returned home, arriving back in London at the age of thirty-four. Although he was able to sell his specimens for a fair amount of money, he was forced to work in his father's hosiery business, a pursuit that left him 'much depressed in health and spirits'. But then he read Charles Darwin's *On the Origin of Species*, published in 1859, and immediately realised the great truth behind it. His own work in the field reinforced Darwin's theories of natural selection and he and Darwin began a correspondence during which Darwin encouraged Bates to write a memoir of his time in the Amazon and even introduced him to his own publisher.

'Among the most beautiful phenomena in nature'

While Henry Bates was exploring the lower Amazon, he came upon something quite unusual. Watching a large group of butterflies in flight, he assumed them to be of one species, merely by the way they looked in the air, but when he caught them, he discovered, to his astonishment, that some of the butterflies were of a different species and were imitating the appearance of the larger group of butterflies.

This resemblance was not readily apparent while they were in the sky— for it was a resemblance of 'external appearance, shapes and colours'—but only when Bates examined the insects closely. He puzzled over this, until he realised that one species of butterfly, a favourite prey of birds, was engaged in mimicry—it had evolved to disguise itself as another species that might be spiny, or bad-tasting, and that birds had learned to avoid.

'These analogies to me appear to be among the most beautiful phenomena in nature,' he wrote. Back in England, he published a paper on the subject entitled 'Contributions to an Insect Fauna of the Amazons Valley. Lepidoptera: Heliconiidae' and it received wide circulation. Charles Darwin was delighted with it because, as Bates himself wrote, 'I believe the case offers a most beautiful proof of the truth of the theory of natural selection.' Darwin generously wrote to Bates: 'I am rejoiced that I passed over the whole subject [of mimicry] in the "Origin", for I should have made a precious mess of it.'

This particular type of natural mimicry became known henceforth as 'Batesian mimicry' and it is a recognised principle of evolutionary science.

When *The Naturalist on the River Amazons* was published in 1863, it was an immediate success and went through several printings. More happiness came to Bates when he married twenty-two-year-old Sarah Ann Mason, a butcher's daughter. Still, it was a continuing problem for Bates that he was unable to find work in his profession. He applied for a job in the Zoological Department of the British Museum, but failed to receive it when the post when to a poet who was far less qualified than Bates but had connections with the Board of Directors. Despite Bates's travels and publications, the class system in England worked against a man who was the son of a humble hosiery manufacturer.

Finally, Bates—in large part through the backing of his publisher, John Murray—was given the post of assistant secretary of the Royal Geographical Society, where he stayed for the next twenty-eight years, helping to keep the society's operations running smoothly and editing several important publications. When he died of bronchitis in 1892, at the age of sixty-seven, he was no longer the 'solitary stranger' of his youth, but a man with a wife and five children (two of whom had become farmers in New Zealand) who had made an extraordinary contribution to nineteenth-century natural science.

MARIANNE NORTH
'a very wild bird'

Like her near contemporary Mary Kingsley (see page 270), Marianne North had quite a tongue on her. At a time when marriage was considered a holy calling for females North called it 'a terrible experiment' that turned a woman into little more than an 'upper servant ... to be scolded if the pickles are not right and then she will have to amuse herself by flirting with the most brainless of the Croquet-Badmintons'—this being the fictional name she bestowed upon the British upper classes, whom she held in disesteem. A child of this same upper class, North escaped her pickle-scolding fate by dint of her flair for painting plants and lust for travel. She was an amateur naturalist—although she did discover five species of flowers—but her vibrant oil paintings (of which over eight hundred hang on the walls of the Marianne North Gallery in the Botanical Gardens at Kew) brought the beauty of exotic plants home to a British audience and turned North into what the flower historian Jane Brown calls 'the high priestess' of the Victorian age of flowers.

'The one idol and friend'
Marianne North's autobiography (defiantly but accurately entitled *Recollections of a Happy Life*) starts in typical fashion: 'It began at

Hastings in 1830, but as I have no recollections of that time, the gap of unreason shall be filled with a short account of my progenitors.' Her progenitors were, in fact, the distinguished North family, particularly her great-great-great-great-grandfather Roger, Lord North, Attorney General under King James II. As a little girl, North was particularly fond of Lord Roger's portrait in the family gallery: 'For Roger I had an especial respect, as the brown curly wig was said to be all his own, and not stuck on with pins driven into his head as my doll's wig was, and I thought he used to look down on me individually with a calm expression of approval.'

Of other ancestors, North did not have such a good opinion: 'My Grandfather Frederick Francis ... married the Rector's daughter, and did nothing to distinguish himself but have the gout, which gave him an excuse for bad temper.' But she dearly loved her father, Frederick, a landowner and Member of Parliament from Hastings, whom she called 'the one idol and friend of my life'. North had a happy childhood in the big family manor, with two sisters and a brother, a donkey named Goblin and an insatiable desire for the novels of James Fennimore Cooper, reading them 'until I fancied myself in the virgin forests of America'.

Known in the family as 'Pop', North in her youth was a bit of a tomboy who took sketching and singing lessons. When she was fifteen years old, in 1845, her long-ailing mother died—'She enjoyed nothing and her life,' North wrote with typical bluntness, 'was a dreary one.' Before dying, her mother made North promise never to leave her father, a promise North was only too happy to keep. She moved to London with him when the House of Commons was in session and kept his flat, and while living with him met a lot of exciting people, including Sir William Hooker, head of the Botanical Gardens at Kew, who gave her a gift of 'a hanging bunch of the *Amherstia nobilis* [the so-called Pride of Burma] one of the grandest flowers in existence,' which made her long to see the tropics.

Hey Diddle Diddle

When Marianne North was a child, her father allowed the artist Edward Lear, who would become the author of The Book of Nonsense *and* The Owl and the Pussycat, *to stay as a lodger in the gardener's cottage on the grounds of the family manor in Hastings while he was at work on some commissioned paintings. Since one of these included a pastoral scene with sheep, Frederick North actually bought a few of the animals and 'kept them in the field within sight of [Lear's] room—a kindness he never forgot'.*

Lear delighted the young Marianne. As she wrote:

> He used to wander into our sitting-room through the windows at dusk when his work was over, sit down at the piano, and sing Tennyson's songs for hours, composing as he went on, and picking out the accompaniments by ear, putting the greatest expression and passion into the most sentimental words. He often set me laughing; then he would say I was not worthy of them, and would continue the intense pathos of expression and gravity of face, while he substituted Hey Diddle Diddle, the Cat and the Fiddle, or some other nonsensical words to the same air.

North commented dryly that she 'was never able to appreciate modern poetry' after these encounters. Lear remained close to both Marianne and her sister Catherine, who married the poet (and secret pederast) John Addington Symonds and had four children by him. A notation in Lear's journal says that while waiting for Symonds in his office one day in 1867, Lear 'drew the Owl and the Pussycat for his children'. He seems to have written the famous nonsense story especially for Catherine's charming daughter, Janet, Marianne North's niece.

For the next fourteen years, Marianne North's life was everything she could have wanted it to be. She spent a great deal of time travelling with her father throughout Europe and the Middle East, both on vacations and for business, and continued to develop her love of plants and flowers and her talent for painting them. At first working in watercolours, she was taught by talented artists—a Dutch painter named Miss van Fowinkel and later by Queen Victoria's own flower painter. Others in her family moved on—her brother married, followed by her sister Catherine, some eight years Marianne's junior— but Marianne stuck with her father. By the 1860s she had started to perfect her technique for painting flowers—working swiftly, now in oils, which she found addictive ('a vice like dram-drinking, almost impossible to leave off once it gets possession').

But then, on 29 October 1869, her father died after a short illness. 'The last words in his mouth,' North wrote, 'were "Come and give me a kiss, Pop, I am only going to sleep".' Her heartbreak palpable even as she was writing her memoir many years later, she added: 'He never woke up again and left me indeed alone.'

'I had long had the dream'

Now the question for North became what to do. She was thirty-nine years old and independently wealthy, and she did not have to share her fortune with a husband. In the months following her father's death, she fled England for Sicily. 'I could not bear to talk of [her father] or of anything else, and resolved to keep out of the way of all friends and relations until I had schooled myself in that cheerfulness which makes life pleasant for those around us.'

Arriving back home to the flat she had kept for her father in Victoria Street, she decided that the thing she needed to do most was to travel: 'I had long had the dream of going to some tropical country to paint its peculiar vegetation on the spot in natural abundant luxuriance,' she wrote. Coincidentally, that summer of 1871, an old American friend, one Mrs S., invited her to travel with her back to the United States. North took her up on it and they embarked together in July. Landing in Boston, North spent several months with her friends, making wry observations about the country—'there are no porters in America', she wrote one day when she was unable to get help as she returned laden with packages from shopping. 'But everyone is courteous and helpful if you are civil too.'

But it was the profusion of plants and flowers that impressed her that summer—the white orchids, yellow sunflowers, glorious pines and dark ferny woods. On a trip to Quebec she was dazzled by the autumn colours ('Nothing but our most brilliant geranium-bed could rival the dazzling variety of reds and crimsons') and she was able to visit some of the spots she had read about in the work of the historian Francis Parkman and in James Fennimore Cooper's novels, stories of French and Indians and the conquest of America.

After a journey through the Midwest, North made her way to Washington and New York, at which point, in December of 1871, she decided to try to realise her dream of painting on a tropical island. She took passage aboard a ship heading to Jamaica and arrived there, 'entirely alone and friendless', on Christmas Eve. In this regard, North was not being entirely frank in her memoirs—she did arrive alone but was not without means, having letters from some of her well-connected friends in England and certain acquaintances who had settled in Jamaica. Through these she rented a house 'amidst the glorious foliage of the long-deserted botanical gardens of the first settlers'—and what habitation could have been more perfect for her?

Her first decision was to buy 40 kilograms (90 pounds) of bananas and hang them from the ceiling of the verandah 'instead of a chandelier'. She ate them from the top downwards as they ripened. And then she began to paint.

'A tangle of all sorts of gay things'

Marianne North's sister wrote about her after her death that she was 'no botanist in the technical sense of the term; her feeling for plants in their beautiful living personality was more like that which we all have for our human friends'. This can be seen in North's description, jotted down in her journal soon after reaching Jamaica, of the vegetation surrounding her hidden little home:

> *The richest foliage closed quite up to the little terrace on which the house stood; bananas, rose-apples (with their white tassel flowers and pretty pink young shoots and leaves), the gigantic bread-fruit, trumpet trees (with great white-lined leaves), star-apples (with brown and gold plush lining to their shiny leaves), the mahogany trees, mangoes, custard apples, and endless others ... A tangle of all sorts of gay things underneath, golden-flowered alamandas, bignonias, and ipomeas over everything, heliotropes, lemon-verbenas, and geraniums from the long-neglected garden running wild like weeds ... over all a giant cotton-tree quite 200 feet [60 metres] high was within sight, standing up like a ghost in its winter nakedness against the forest of evergreen trees ...*

She was, she wrote, 'in a state of ecstasy'. She got up at dawn, painted as fast as possible on an easel she had set up among the wild gardens, and then, when the afternoon rains came, she worked indoors. People—English exiles living on the island—tended to 'find her out' and invite her to stay with them, but she refused these visits or cut them short, or begged off the long, formal breakfasts so she could get up at dawn and paint. Understanding hosts would merely send her out tea. Her Jamaica paintings, her first major body of work, are bright, bold and colourful, deep with feeling, if not always as precisely exact as the work of Maria Sibylla Merian (see page 22). She painted 'a slender tree-fern with leaves like lace-work, rising out of a bank of creeping bracken, which carpeted the ground and ran up all the banks and trees, with marvellous apple-green hue'. She captured a mahogany tree, 'the hardest and blackest wood on the island ... its velvety leaves and trumpet-flowers of copper and brass tints made a fine study'.

North stayed in Jamaica until June of 1872, travelling widely over the island and producing dozens of paintings, after which she headed back to London.

'All strung with crystal beads'

Arriving back in England, she wrote to her friend, the traveller and writer Amelia Edwards, that she intended to visit her father's sister in Hastings, a journey that would bring up memories that, in any event, were never far from the surface, no matter where she went: 'It will be a sad visit, but I have a sort of longing to touch my darlings grave again it is all I care for there—or indeed anywhere—and in all truth Amy I have no love to give you or anyone—it is all gone with him—it would be untrue to pretend otherwise ...'

Despite her travel and growing pride in herself as a painter, North was literally haunted by the ghost of her father ('sometimes as I sit and paint he seems to come and watch me, and the very thought of him keeps me warm'). Still, determined 'to devote myself to painting from nature and try to learn from that lovely world which surrounded me ... how to make that work henceforth the master of my life,' she decided to set out for even more exotic climes than Jamaica: Brazil.

Marianne North landed in Brazil, in the province of Pernambuco, in late August 1872 and immediately set out into the jungle. Although she did carry letters of introduction to smooth her way, she followed her usual practice of avoiding the 'Croquet-Badmintons' of the world. Suzanne Le-May Sheffield writes in her book *Revealing New Worlds: Three Victorian Woman Naturalists*, North 'preferred to live, travel and work outside of the confines of established Empire oases. In fact, her work seemed to conveniently necessitate it.' In other words, painting in remote spots appealed to her inclination to be left alone and facilitated that desire at the same time.

In Brazil, after a short time in Rio, she joined a mule caravan and travelled deep into the highlands during the rainy season, over muddy roads and treacherous 'corduroy' bridges made of round logs, which often slipped apart, allowing the unwary traveller or mule to fall through up to the legs. But when North reached the high plains she found the flowers beautiful—'campanulas of different tints, peas, mallows, ipomoeas, velvety stalks and leaves; many small tigridias, iris, and gladioli, besides all types of sweet herbs'. Near the hut of a freed black slave 'were masses of bright scarlet-flowering euphorbia, from the juice of which the Indians poison their arrows'.

Finally, after a journey that left her with 'blistered lips and slightly browned hands', she ended up in the small town of Morro Velho, where she settled in to paint in a house surrounded by jungle. It seems that the profusion of life revived her. She took long walks with a mongrel dog that adopted her and warned her of the approach of poisonous snakes, and she was quite an object of curiosity to local girls, who would approach her verandah, sit down and poke their legs through the balustrades. When she threatened to paint their toes— considered a 'disgrace'—they would dart off, but slowly return later.

Seeing that North liked to interject insects and other small wildlife into her botanical scenes, the locals began 'collecting' for her, and soon she found herself with many types of insect specimens— praying mantises, stinging caterpillars, scorpions—which she had to handle quite carefully. (They were, in any event, hard to preserve—she would hang them on a string overnight but find them eaten by bats in the morning.) Yet, with all the newness and strangeness around her, North settled in to paint, producing some of the classic work that would later hang in the Marianne North Gallery at Kew, such as *Macrosiphonia longiflora*: 'like a giant primrose of rice-paper with a throat three inches [75 millimetres] long; it is mounted on a slender stalk and had leaves of white plush ... and a most delicate scent of cloves'. Whenever possible, North still painted in the quiet of the morning, near dawn: 'How lovely those wet mornings were! And huge spider webs, all strung with crystal beads, so strong that they seemed to cut one's face riding through them.'

'Perfect world of wonders'
North finally returned to England, arriving in September 1873, having completed a hundred gloriously colourful paintings. In the winter of 1875, she was off again, travelling to Tenerife. She was constantly

in motion and saw beauty everywhere. In Tenerife, 'the ground was white with fallen orange and lemon petals ... I never smelt roses so sweet'. Arriving back in London briefly, she set forth again, this time to Japan, Singapore and Java, crossing America first. Once again, she found the locals amusing—in Chicago, one said to her: 'You're a-going to paint pictures of Japan, are you? Wall! I wish you success; I should like to go along, too!'

North gathered dust in her teeth as she crossed the country via a bone-jolting train ride. After a stop taking in the wonders of Yosemite Valley, she arrived in San Francisco and on 16 October sailed to Japan aboard a luxury liner, the *Oceanic*. She arrived on 7 November, after watching the sun rise out of the sea and 'redden the top' of Mount Fujiyama. Helped by members of the British legation, North found herself a room with a view of the mountainous horizon near Kobe and began to paint, but she did not fare so well in Japan. The people she considered to be 'like little children ... and have scarcely any natural affection towards one another'. (She was offended that Japanese mothers did not pick up and comfort crying babies.) She disliked the way men spoke to people they considered their inferiors, using 'a most guttural tone of voice. I used to fancy strangulation must ensue after much of it.' Whenever she sat out to paint, she was surrounded by crowds of people, very polite, but staring at her.

After a bout of rheumatic fever, it was a relief for her to leave and head for Hong Kong and then Singapore, where she landed in January of 1876. She visited the Botanical Garden there and was quite impressed, as she was by the jungle that rose everywhere beyond the confines of the city. 'I found real pitcher-plants (Nepenthes) winding themselves among the tropical bracken. It was the first time I had seen them growing wild, and I screamed with delight.' She loved the gardens of the English friends she visited, vivid with colour, rare orchids hanging everywhere, 'scores of *phaloenopsis* in full flower, like strings of white butterfly, hovering in the air with every breath of air'. (Always scornful of affectation, however, she disliked the garden of a

Chinese man whose many plants 'were cut into absurd imitations of human figures and animals, to me highly objectionable'.)

Before returning home, North made stops in Borneo, Java and Ceylon (Sri Lanka). During all the time she spent in these tropical locales, 'with everyone going regularly to bed in dark rooms' during the heat of the day, she refused to. The daylight hours were best for painting and that was what she was there for, visiting this 'perfect world of wonders', surrounding herself with the beauty, and comfort, of flowers.

'A very wild bird'

Marianne North returned to England in February of 1877 with another hundred or so paintings, her total output now at about five hundred. She was becoming a bit of a celebrity in London. The newspapers had begun to cover her travels and her paintings were causing quite a stir, for their bold colours and because they had been done in nature, with North often setting the scene with insects or other creatures. In a day before colour photography, North's paintings were powerful images of the glories of nature. But North had other goals aside from aesthetic ones. As Suzanne Le-May Sheffield points out, one of the reasons North was driven to do what she did was because she wished to educate the public. 'I found people in general woefully ignorant of natural history,' North wrote in *Recollections of a Happy Life*. 'Nine out of ten of people to whom I showed my drawings [thought] that cocoa was made from coconuts.' It drove her crazy that people admired her paintings even as they held them upside-down. She wanted the public not only to see the beauty she saw, but to actually know what they were looking at.

To North's great pleasure, the Kensington Museum decided to exhibit all five hundred of her paintings and she spent a good deal of time before her next trip putting together the catalogue, adding in as much information as possible for the edification of the unenlightened. Then she was off again, in September, this time for fifteen months in India, but first she made her will, leaving her paintings to the Kensington Museum 'to do with them as they like, for the good of others'.

North found India fascinating, but also 'full of darkness and uncanniness', as she described the interior of one temple she entered. 'Starvation, floods, and fever' were all around—but so was extraordinary beauty. Looking out from her room in Bombay, she described the sun coming up 'like a round red ball behind the purple hills and hanging smoke of the city'. She was so taken by the Taj Mahal that it was 'some days' before she felt that she could approach it. Obviously, she was in her element. Visiting the hill country of India, she painted *Rhododendron arboreum*, with the majestic Mount Kangchenjunga behind it and her favourite dawn sun lighting it all (the painting took her a week of dawn rising outside Darjeeling). When she visited Simla, in northern India, she tramped through 'the most glorious forests of Smithina pine' and did a brilliant painting of *Rhododendron dalhousiae*, which must have pleased Sir Joseph Hooker (see page 198), since he had named it.

These were glorious days for North. 'I am a very wild bird and like liberty,' she wrote. She came back to London in 1880 with two hundred more paintings.

'Wander for miles and miles'

North now had another showing of her work, this time at a gallery on Conduit Street, in London. A local newspaper suggested that her botanicals might be perfect on display at Kew Gardens, now directed by Sir Joseph Hooker. She thought about this for a while, and then—having missed a train she was supposed to catch—decided to use the time to write to Hooker and ask 'if he would like me to give [my paintings] to Kew Gardens, and to build a gallery to put them in'.

To her great pleasure, Sir Joseph wrote back and answered in the affirmative, although he rejected her suggestion that refreshments be served. With typical thoroughness, North sprang into action. She asked the well-known architect James Fergusson to design the building on a spot she had chosen herself—'far from the usual entrance-gates'. She was also able to have a small artist's studio built there, for her use or that of another artist, a 'quiet room in the gardens in which a specimen could be copied'.

Coffee, tea and art

When Marianne North first envisioned what became the Marianne North Gallery at Kew Gardens, she wrote to Sir Joseph Hooker that she 'wished to combine this gallery with a rest-house and a place where refreshments could be had—tea, coffee, etc.' Hooker turned her down, explaining that the Kew Gardens received, at that time, seventy-seven thousand visitors a year and that to provide them all with tea was impossible, especially should they come, as North wrote with her usual sly humour, 'all at once on a Bank Holiday'. 'He mentioned too,' North added, driving home the dagger, 'the difficulty of keeping the British Public in order'.

North, however, did not give up easily. She knew from her travels how welcome refreshment was after a long day of sightseeing, no matter how pleasurable. And she also felt that Hooker was being somewhat stuffy, since he also insisted that many of his visitors were scholars who simply did not need tea as they viewed rare botanicals through their pince-nez. However, she was unable to change his mind and, with the help of architect James Fergusson, set about designing the gallery and filling it with her pictures. On the day it opened, in 1882, visitors noted to their puzzlement that above the doors North herself had painted pictures of tea and coffee plants, a kind of joke on Hooker.

In the twentieth century, the North Gallery fell into some disuse, with few people who passed it on their tour of Kew Gardens understanding what the temple-like building was. However, much of North's work has been restored and the North Gallery reopened again in 2009, some one hundred years after it was first opened to visitors—this time serving coffee, tea and biscuits. Marianne North had the last word, after all.

One day, as she was arranging all this, she was introduced to Charles Darwin through his daughter. Darwin was, as she wrote, 'the greatest man living', and she was thrilled that he wanted to see her and had heard of her. He had one suggestion for her—to go to Australia to paint the vegetation there 'which was so unlike that of any other country'. Almost straightaway, North did. She arrived in Brisbane, stayed at an unattractive hotel that was 'overfull', and for once did not shun an invitation from upper class English friends, accepting an offer to stay at Government House, which had the added advantage of abutting the Botanical Gardens. North's fame had finally begun preceding her wherever she went—she recorded that the local paper had written an article of welcome, and a little boy had approached her to ask if she would be knighted by the queen, something she found quite amusing. (In fact, a male naturalist of North's quality would certainly have been knighted.)

As usual, as soon as possible, North left the urban area and ventured out into the countryside, fascinated by Queensland and by the kangaroos she saw ('perhaps twenty ... all hopping down the hill in single file'). One day, she wrote, she spread her handkerchief under a eucalyptus tree 'and lay on my back examining the Eucalyptus leaves overhead for an hour at least ... It is a libel to call them shadeless trees. Just at noon the knife-edges of the leaves are turned towards the sun, thinking more of keeping themselves cool by exposing the least possible surface to the sun's rays, than of shading the ground below.'

After spending a short time in New South Wales, North headed to Western Australia, where she met 'Mrs. R.', the great Australian naturalist painter Ellis Rowan, who 'introduced me to quantities of the most lovely flowers—flowers such as I had never seen or even dreamed of before' and which Rowan painted 'in her own exquisite way' on her trademark grey paper. The flowers in the garden of the hotel North was staying in literally left her breathless: 'In one place I sat down, and without moving could pick up twenty-five different flowers within reach of my hand ... The whole country was a natural flower garden, and one could wander for miles and miles among the bushes and never meet a soul.'

'I am such an old vagabond'

Leaving Australia, North found her way home with dozens of paintings. One of the first things she did was to pay a visit to Charles Darwin, and she took some of her pictures with her. He wrote her a gracious note in August 1881 that she treasured ever after. It read, in part: 'I am so glad that I have seen your Australian pictures ... to the present time, I am often able to call up with considerable vividness scenes in various countries which I have seen ... but my mind in this respect must be a mere barren waste compared with your mind.'

In 1882 the Marianne North Gallery at the Botanical Gardens at Kew had its grand opening. North was fifty-two years old and her health was beginning to fail. The building she had Fergusson design resembled a Greek temple, with high windows that cast a lofty light in the rooms (accounting for the fact that North's paintings have suffered comparatively little sun damage). The lower part of the gallery walls were made of some 246 varieties of wood that North had brought back from all over the world. On the walls of the gallery were hung 832 of North's paintings, packed closely together (one observer said the gallery resembled a gigantic postage stamp album). But what a grand tour of the world's flowers it was, beginning in America, moving on to Japan, India, Java, Australia, Jamaica and South America.

North would travel to other places in the years before she died in 1890—to the Seychelles, to Africa and back to South America—but the Marianne North Gallery was her crowning achievement, the one place her startlingly lifelike, yet profoundly emotional paintings— the plants with their 'beautiful living personality', as North's sister wrote—could be seen in their entirety and capture her life as she had lived it. Once it was all hanging on the walls, Marianne North, having done her part for society, could go back to the solitude she so treasured, and which she shared with the shade of her father.

'I am such an old vagabond that I own to be delighted to be perfectly free again', she wrote during one trip to Italy. 'Staying with no one, having no fixed dates for going anywhere and not even a servant to dog my footsteps. I sat on the bench at the top of the hill and waited for the clouds to roll their way upwards and thought with glee—there is no reason except hunger which need drive me down to the lake again—for hours to come—it was so grand there ...'

'A fierce intense vitality'

During her visit to Australia, Marianne North met and befriended Ellis Rowan, a painter and naturalist who was, in some ways, extraordinarily like her. The petite Rowan travelled all over Australia, as well as to New Zealand, New Guinea, the Himalayas, Europe, the United States and the Caribbean, ultimately painting over three thousand botanicals in the field, under all sorts of trying conditions.

Rowan was born in Melbourne in 1848, the eldest of seven children. Her maternal grandfather was an English naturalist named John Cotton; Ellis was only an infant when he died, but her mother preserved his sketchbooks for her. Ellis received a proper Victorian upbringing, much like Marianne North, with private art and music lessons. In 1869, when she was twenty-one, she took a trip to England, where she may have been introduced into London's artistic circles (Rowan was always vague about this) by her uncle Sir Charles Eastlake, who was well acquainted with London's artistic elite. Whether she met them or not, she returned home to win a bronze medal at the 1872 Intercolonial Exhibition in Melbourne for flowers she drew on a four-panel screen.

Soon after this, in the early 1870s, Rowan's father bought a large parcel of land near Melbourne and, being an amateur botanist, filled it with flowers both familiar and exotic. And Ellis Rowan began to paint them, sending her work to her father's friend Ferdinand von Mueller, head of the Royal Botanical Gardens in Melbourne. Unlike Marianne North, Rowan married and had a child, but she continued to paint flowers, gradually gaining renown. In 1880 she met North on the south coast of Western Australia and, as she wrote, 'I became her devoted admirer and she became the pioneer of my ambition.' North, for her part, 'admired [Rowan] a lot for her genius and prettiness, she was like a charming spoiled child'.

But Rowan was much more than that. By now forty years old, her work matured as she travelled, and by the end of the 1880s she had become the most famous painter in Australia, much to the chagrin of male colleagues who attempted to slander her whenever possible. There was much resentment of the fact that she left her husband and son at home while she travelled, thus turning traditional mores on their head, but Rowan's marriage was a good one, and she mourned her husband when he died prematurely in 1892. After the death of her twenty-two-year-old son during the Boer War, Rowan continued to travel, painting butterflies (she claimed 2175 of them) as well as stunning plant studies. She died in 1922, 'a fiery, intense vitality,' as one memorial had it, 'in a seemingly most fragile personality'.

FREDERICK COURTENEY SELOUS
'the last of the big-game hunters'

Frederick Courteney Selous was a man of contradictions, one foot in real life, the other in legend. On the one hand, he was such a famous African hunter that the author H. Rider Haggard based his celebrated character Allan Quatermain on him. Selous, like Quartermain, was a crack shot, had countless scrapes with death, was, even into his sixties, more physically powerful than most twenty-year-olds—and was the soul of modesty, to boot. American President Theodore Roosevelt, a good friend of Selous, wrote in 1908 that 'Mr. Selous is the last of the big-game hunters of South Africa'. This was even then an exaggeration but it is true in that Selous and his fellows hunted so much that they helped wipe out populations of elephants and other animals in numerous wild regions of South Africa. 'If I can't get good shooting in this world, I'll have it in the next,' Selous vowed passionately as a young man.

This would tempt us to write him off as merely one of the many nineteenth-century butchers of big game, in it only for the trophies and the body count, but this would not be an entirely fair portrait. For Selous was a naturalist as well—to be sure, not an early twenty-first century type of naturalist, with a motto of 'first do no harm', but a concerned lover of wilderness who collected butterflies between lulls in battles with German troops during World War I and wrote highly detailed portraits of the people and animals he encountered in over forty years in the wild. Roosevelt—himself cut from the same cloth—recognised this. Continuing his description of Selous, the president observed: 'Mr. Selous is much more than a mere big-game hunter, however; he is by instinct a keen field naturalist, an observer with a power of seeing and remembering what he has seen.'

'One day I am going to be a hunter'

Frederick Selous was born in Regents Park, London, on 31 December 1851, to well-off and aristocratic parents. He was the descendant of what Selous's biographer John Guille Millais called 'an ancient and artistic race' of French Huguenots who had been exiled in England since the French persecution of the Huguenots. Selous's father, Frederick, head of the London Stock Exchange, still mourned the St Bartholomew's Day massacre of nearly two centuries before. Even as he made money hand over fist, he developed a reputation as one of the finest amateur clarinet players in England. Selous's mother, Ann, was a published poet and his younger brother Edmund would go on to become a famous ornithologist (who abhorred hunting).

From the very beginning, Selous was a bit of a wild child. Of average height and build, with arrestingly limpid blue eyes, he skirmished continuously with authority figures, especially gamekeepers whose domain he loved to invade with small calibre rifles. When he was nine years old, the headmaster of his school found him lying on the bare floor of his dormitory room one night, clad only in a nightshirt. When asked why he was not in bed, Selous replied: 'Well, you see, one day I am going to be a hunter in Africa and I am just hardening myself to sleep on the ground.'

When he was fourteen, Selous went off to school at Rugby, where he read avidly, particularly the work of the African explorer David Livingstone, and spent as much time as possible outdoors. In January of 1867, shortly after his sixteenth birthday, he was home from school on holiday and displayed what would become characteristic Selous luck by escaping from the horrible Regent's Park skating disaster. Thousands of Londoners had been coming daily to skate on the park's huge ornamental lake, but a thaw had caused the ice to thin. Ignoring warnings, Selous and others went out skating on the lake. Late in the afternoon, Selous was on a central part of the water, near a small island, when he heard a terrible cracking sound and turned to see jagged cracks shooting across the lake. As the ice began to break up, and people, screaming, plunged into the frigid water wearing their winter clothing, Selous kept his head, lay down flat, and crawled from ice floe to ice floe before finally making it to dry land. Forty-nine Londoners died that day.

Frederick Selous was studying to be a doctor, and when he was seventeen he was sent to Germany to further these studies, but those who knew him were in little doubt that he could not be cooped up too long in an office—particularly after he was forced to flee Germany in the wake of knocking a gamekeeper unconscious during a fracas over some bird eggs he had pocketed.

While he was in Germany, however, the young man had met numerous people with connections to Africa, particularly southern Africa, which had a large German population, and he prevailed on his parents to send him there. On 4 September 1871, at the age of nineteen, he landed at the Cape of Good Hope with four hundred pounds in his pocket and a burning desire to make his living as an elephant hunter.

'Why, you're only a boy'

The first thing Selous did was travel overland to Matabeleland, an area in what is today western Zimbabwe. He arrived after a journey of two months and, without any experience at all in big game hunting, set off into the wild. Right away, he began having the adventures he was known for. Shortly after entering the bush to begin hunting, Selous and his party encountered a herd of about twenty giraffes and immediately gave chase on horseback. Selous accidentally crashed into a tree and nearly broke his leg. When he recovered from being stunned, he found himself completely alone in the forest, with no water and with only three cartridges left. He wandered for two days with no idea where he was. Climbing a hill, he stood on a heap of rocks and observed 'a vast ocean of forest, as far as the eye could reach'. Awakening after his third night in the wild—with no water or shelter—he found that his horse had wandered off during the night. Growing steadily weaker and weaker, he walked on and on through the forest. It was in a state of near collapse that he met two native cattle herders, who rescued him.

What Selous was really after, however, was elephants, whose ivory tusks could be sold for a nice price to traders. In order to achieve this, Selous wrangled an introduction to Lobengula, king of the Matabele people, and begged him for permission to hunt elephants in his territory. At first Lobengula, seeing Selous's youth, laughed at him. 'Was it not steinbucks [a small antelope] that you came to hunt? Why, you're only a boy.'

Despite Selous's protestations, Lobengula got up to leave, a scene nicely described in Selous's book *A Hunter's Wanderings in Africa*:

> *[Lobengula] then rose without giving me any answer. He was attended by about fifty natives who had all been squatting in a semicircle during the interview, but all of whom, immediately he rose to go, cried out, 'How! How!' in a tone of intense surprise, as if some lovely apparition had burst upon their view; then, as he passed, they followed, crouching down and crying out 'Oh thou prince of princes! thou black one! thou calf of the black cow! thou black elephant!'*

Despite this response from the king, Selous persisted and was able to see him again a few days later. This time Lobengula was willing to talk. Had Selous even seen an elephant before, the king asked. Selous had to reply that he had not. At this the king sighed and gave him permission to hunt anywhere he liked, but avowed that the huge creatures 'would soon drive you out of the country'.

A near miss

Far from being driven out of the country, Selous became one of the finest elephant hunters in Africa. This was despite many miscalculations. One of his initial problems was his guns. On his journey to Matabeleland, he had purchased two powerful elephant guns from a local trader. They weighed only 2.7 kilograms (6 pounds) apiece and so were quite light, and thus, Selous wrote, he was able to load them on the run simply by dipping his hand into a bag of powder that he kept at his side, and then putting in a 120-gram (4-ounce) iron ball. However, the guns 'kicked most frightfully, and in my case the punishment I received from these guns has affected my nerves to such an extent as to have materially influenced my shooting, and I am heartily sorry I ever had anything to do with them'. (However, between 1872 and 1874 Selous would kill seventy-eight elephants with these guns, testament to his skills as a hunter.)

Soon after gaining permission from Lobengula, Selous journeyed into the so-called 'fly country' (a part of Matabeleland where the tsetse fly was particularly pernicious), on foot, with three Ndebele bearers. Following trails of broken trees and bent bushes, they came upon an old bull elephant standing by himself in a grove of trees. He was unaware that Selous and the others were there. At the behest of the Ndebele, Selous took off his pants and was now wearing just cotton shoes, hat and shirt—'nice light running order', as he wrote. Shouldering his weapon, he fired a single round into the elephant from close range. The elephant roared and took off and Selous followed, finally killing it with a few more rounds. That night, he dined on 'elephant's heart cooked on a forked stick over the ashes' of his fire and considered himself to be a happy man. The animal's tusks together weighed over 45 kilograms (100 pounds).

Not all of his elephant kills were that easy, however. One of Selous's closest calls came some time later, when he was hunting elephants on horseback. He had attacked a herd of the giant creatures and shot four of them. He picked out, for his fifth victim, 'a good cow', which he shot in the usual place, behind the shoulder, aiming for her heart. Instead of falling to the ground, the animal merely snorted and then charged Selous, who was in the act of reloading and could not fire. Instead, he turned his horse to run:

> I heard two short sharp screams above my head, and had just time to think it was all over with me when, horse and all, I was dashed to the ground. For a few seconds I was half-stunned by the violence of the shock, and the first thing I became aware of was a very strong smell of elephant. At the same instant I felt that I was still unhurt and that, although in an unpleasant predicament, I still had a chance for life. I was, however, pressed down on the ground in such a way that I could not extricate my head. At last with violent effort, I wrenched myself loose, and threw my body over sideways ... As I did so I saw the hind legs of the elephant, standing like two pillars before me, and at once grasped the situation. She was on her knees, with her head and tusks in the ground, and I had been pressed down under her chest, but luckily behind her forelegs.

Frantically, Selous clawed his way out from under the wounded elephant and staggered to the shelter of a nearby bush. The elephant attempted to look for him, but grew tired and simply walked off. Selous was quite lucky to be alive. A little later, an old elephant attacked one of his porters. Since the animal's tusks were worn down, he could not gore his victim and so simply held the man down with one giant foot and tore him apart with his trunk, piece by piece.

Lions, buffalos and elephants

One of the topics of debate among the great African hunters of Selous's time was which animal was the most dangerous to hunt—the lion, the buffalo or the elephant.

The hunter William Finaughty, arriving in Africa at the age of twenty-one in 1864, travelled to Matabeleland at a time when it was ruled by the brother of the great Shaka Zulu. Finaughty found a land filled with tens of thousands of animals and once watched a great dance in which 2500 warriors took part and for which 540 oxen were slaughtered. Elephants soon became his specialty—within two months in 1868 he shot ninety-five elephants, whose ivory weighed over 2200 kilograms (5000 pounds). Despite being nearly killed by a bull elephant, Finaughty considered the wild buffalo by far the most dangerous game. A charging buffalo was almost impossible to bring down, even with a direct hit by a large calibre bullet. And wounded buffalos had the nasty habit of circling behind the searching hunter and attacking him from the rear. 'Far better,' wrote Finaughty, 'to follow up a wounded lion than a wounded buffalo ... A man who is out after buffalo must shoot to kill and not to wound.'

The Boer hunter Piet Jacobs, the most famous hunter in South Africa at the time Selous arrived, is supposed to have killed four to five hundred elephants and numerous lions—and he was once ferociously mauled by a lion. He swore, however, that elephants were the most dangerous, because of their massive size and because they were intelligent creatures.

But Frederick Selous, who had close encounters with both elephants and lions, was certain that lions were Africa's most dangerous creatures. Generally, he avoided killing lions—he killed only thirty-one in all the years he spent in Africa, far fewer than either Finaughty or Jacobs—but he carefully studied the animals. They were the most dangerous because of the way they turned humans into prey—attacking mainly in darkness, coming in low and fast through the grass, sinking their massive jaws into the throat. They were unpredictable, especially when hungry, and, at about 200 kilograms (450 pounds), the fastest, most powerful animals of their size. They possessed, wrote Selous, 'two requisites for terrestrial happiness—good appetite and no conscience'.

Lions

For the next eighteen years, taking time out for brief trips back to England, Frederick Selous hunted in South Africa, making his living harvesting ivory from the elephants he shot but also building up the collection of five hundred specimens—everything from lions to butterflies—that he would eventually bequeath to the Natural History Museum in London. A man of prodigious energy, Selous would write his experiences down in journals, which he would later flesh out in his books.

Selous made an especially close study of lions, which he considered the human's most worthy foe in the African bush. One day, he wrote, he was galloping his horse through the veldt and suddenly realised that a lion was lying in the tall grass only 20 metres (20 yards) away. Pulling the horse up short, he tried to bring his sights to bear on the lion's head, but the horse was heaving and bucking.

> *As I raised my rifle and looked down the barrel to align the sights upon its head, I saw the black tuft of hair at the end of its tail flicking lightly from side to side, and the forepaws, that had been stretched out straight beyond its nose, drawn slowly under its breast, without the head or body being perceptibly raised. I knew the lion was on the very point of charging ... Just as I saw the crouching beasts' hindquarters quivering, or rather moving gently from side to side, I fired, and luckily my bullet struck it just between the eyes, and crashing into its brain, killed it instantly ... Since that time I have on several occasions watched a cat when stalking a bird go through every movement made by that lion—the same apparently involuntary twitching of just the end of the tail, the same drawing-in of the forepaws beneath the chest, and then the wavy movement of the loins just before the final rush.*

Selous had watched lions disembowel their prey and could report on exactly where they started their meal—either where the thigh meets the stomach, where the skin is thinnest and can easily be ripped with a claw, or at the buttocks, whose rich fat the lions tore off in great gobbets. He even made a study of the lion's mane. Hunters had a tendency to prize lions' manes for how full and dark they were— the richer and blacker, the more prized—and it was thought that different types of manes meant different subspecies of lions. After studying lion siblings, Selous was able to report convincingly that mane fullness and colour was a product of environment—the hotter the area, the thinner and lighter the mane.

Selous's observations were not entirely in the interest of pure science. He felt there were other hunters better at tracking and shooting, but his edge was that he had made a full study of the animals in the area he hunted. As Theodore Roosevelt wrote: 'Probably no other hunter who has ever lived has combined Selous' experience with his skill as a hunter and his power of accurate observation and narration.'

'The present history would never have been written'
By 1890 Selous had come fully into his own as a master hunter and explorer, having traversed much of what is now Zimbabwe and published a best-selling book, *A Hunter's Wanderings in Africa*, about his experiences. Balding, with a neat beard, wearing his characteristic slouch hat pushed back on his head, Selous had the build of an athlete—a whipcord frame and amazing endurance. The Africans he met considered him the best European runner they had ever seen— it was nothing for him to run kilometres through the veldt if the occasion called for it. He ate modestly and drank little alcohol—his chief vice seemed to be tea, which he drank copiously. His fame had grown to such an extent that Europe's finest gunsmiths sent him their newest rifles, hoping he would endorse them.

There is evidence that, in his late thirties, Selous tried to settle down—he attempted to start an ostrich farm, but the job bored him and he returned to hunting. When not felling big game, he collected butterflies. He liked to hunt for them during the hot afternoons, when both humans and animals were sleeping, and he found enough new species to be able to present a very respectable collection to the Cape Town Museum.

In 1890 Selous turned his attention away from hunting to empire building, in the service of the wealthy imperialist Cecil Rhodes. It was Rhodes's dream that British colonies in Africa be joined by a Cape to Cairo railroad, as well as a similar road. Selous enthusiastically endorsed Rhodes's view and decided to help guide a pioneer group of road builders across Matabeleland, a distance of some 740 kilometres (460 miles) through wilderness, much of it quite desolate. Aside from the hardships inherent in this enterprise, Selous and his companions had to deal with the fact that the Matabele, under his own friend King Lobengula, were beginning to resent European intrusions.

By 1892, due in no small measure to Selous's labours as guide, the British had opened up much of the country to road, telegraph and railroad. However, in 1893 what became known as the First Matabele War broke out between the forces of Lobengula and the British. Selous joined the British forces and marched on Bulawayo, Lobengula's capital city. It was on this march that Selous was wounded in action by Matabele fire (the bullet struck him 75 millimetres (3 inches) below his right breast, but fortunately ran around his ribs and exited on the other side, rather than piercing his heart). His luck had held once again.

A recuperating Selous headed back to Britain, where the forty-two-year-old confirmed bachelor surprised everyone by marrying twenty-two-year-old Gladys Maddy. In the longest break Selous had yet taken from Africa, he and his wife travelled extensively through Europe and the Middle East and then returned to Britain, where Selous indulged himself in seal hunting in Scotland. Finally, however, claiming that life in Britain was too expensive, he and Gladys headed back to Africa. They had arrived in time for yet another war with Lobengula's people, the Second Matabele War, in which Selous once again played a prominent part. In one more extraordinarily narrow escape, his horse threw him while he was surrounded by Matabele warriors. As he later wrote:

> *The [Matabele] thought they had got me, and commenced to shout out encouragingly to one another and also to make a kind of hissing noise, like the word 'jee' long drawn out. All this time I was running as hard as I could. As I ran, carrying my rifle at the trail, I felt in my bandolier for how many cartridges were still at my disposal, and found that I had fired away all but two of the thirty I had come out with ... Glancing behind me, I saw that the foremost [warriors] were gaining on me fast, though had this incident occurred in 1876 instead of 1896, with the start I had got I would have run away from them.*

Fortunately, a companion came on horseback to rescue him or, as Selous put it, 'the present history would never have been written'.

'An old wolf'

Between 1896 and 1907 Selous travelled a great deal, hunting in Turkey, the American Rocky Mountains, Alaska, Canada and Newfoundland. During this time, the peripatetic Selous perfected his skills at egg hunting (the pastime that had caused him to flee Germany as a youth). A member of the British Ornithologists' Union, Selous wandered over large, wild parts of England, Scotland, the Orkneys, Turkey, Spain, Hungary, Holland and Iceland collecting eggs. It was not so academic a hobby as it might seem. In Turkey, he took the eggs of the black vulture, griffon vulture and short-toed eagle, attainable only by climbing at some peril up rocky crags.

During this period, Selous became a good friend of Theodore Roosevelt, who was assistant secretary of the US Navy in the late 1890s, and went on to become vice-president of the United States. When William McKinley was assassinated in 1901, Roosevelt became president of the United States, and Selous had a friend in the White

House. He and Gladys visited there in 1903, and Selous spent a fair amount of time rock-climbing with the president, as well as swimming in the Potomac River with him. At night, Selous entertained Roosevelt and his six children with tales of African adventures, tales that fired the president's already strong desire to go on African safari.

In 1908, as Roosevelt prepared to leave office, he turned to Selous for help in planning a year-long African safari. While Selous did not accompany Roosevelt for most of the 1909–10 safari, he did plan the journey—which went from British East Africa, through the Congo and into the Sudan and points north. In fact, as the author Bartle Bull points out, Selous was the first to make the transition from hunter to safari guide. 'Specifically, his organization of the Roosevelt safari launched a booming trade in American safari clients, which became the financial core of the safari business.'

When World War I broke out, Selous had semi-retired to Surrey, where he lived with his wife and their two teenage sons, but even at sixty-three he was in far better shape than men much younger. The minute hostilities began, Selous decided to go back to Africa to fight with the British forces against the German troops stationed in German East Africa. At first the British government insisted Selous was too old, but they finally gave in after he had a thorough physical examination. Selous was made intelligence officer in the Legion of Frontiersmen, a special and flamboyant unit of fighters made up of British hunters and settlers, veterans of the French Foreign Legion, American cowboys, and others.

The fighting was brutal, as Selous and his soldiers encountered seasoned German soldiers leading forces of well-trained native troops. Still, Selous was outstanding. He led an amphibious attack across Lake Nyanza and led men through bush country under conditions that would daunt younger men. When not hunting Germans, marching or fighting, Selous was collecting butterflies with a net he carried with him everywhere. In September 1916, at the age of sixty-five, Selous was awarded a DSO for 'conspicuous gallantry, resource and endurance'.

Early in January of 1917, Selous was leading forces attempting to encircle a retreating German company near the Rufiji River when he and his men came under fire from hidden positions. Raising his head, he placed his binoculars to his eyes to try to find the source of the fire and was hit in the head by a sniper's bullet. He died instantly.

Selous was mourned by a great many people. Not just his wife and children (his older son, an aviator with the RAF, was to die exactly a year later on the Western Front, shot down by German fire), but all the people with whom he had come in contact. He was a man who could run down on elephant, but at the same time write a study of the African waterbuck, complete with commentary on its habits, with its different names in Latin, Afrikaans, English and seven African languages. He was, as Theodore Roosevelt was to write, 'an old wolf' and his like would not be seen again in Africa.

Shooting in London

A story Frederick Selous loved to tell about himself concerned the time that he had ordered a new, high-velocity rifle from an exclusive London gun maker but it did not arrive until an hour before he was due to leave for Africa on his latest hunt. Room in Selous's luggage was limited and he did not want to take a rifle to Africa without testing it carefully, but as he was in Regent's Park, in the middle of London, this seemed impossible.

Until he came up with an idea, that is. He told the maid to call a cab and have the cabman load his luggage. When that was done, he went up to his bedroom window and sighted the new rifle on a chimneystack some 100 metres (100 yards) away. Then he fired five shots in rapid succession. Checking carefully with field glasses, he saw that the shots had fallen in a satisfactory pattern on the brick chimney, indicating that the rifle was a good one. He then cleaned it, put it in its case, donned his hat and went out the front door.

The shots had aroused great curiosity and alarm in the neighbourhood and he found a group of people talking frantically to a police officer about their origin. When asked if he had heard shots, Selous replied that indeed he had. And then he went on his way.

EUGÈNE DUBOIS
Java Man

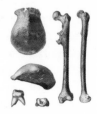

Many scientific travellers of the nineteenth century showed a singular devotion to their craft, braving hardships of all kinds to bring home rare and valuable specimens. But Eugène Dubois's commitment to his finds trumped all of them. In the summer of 1895, as he crossed the Indian Ocean on his way from Batavia (in Java) to Marseilles, a storm arose that threatened the safety of the steamer on which he had booked passage for himself, his wife, Anna, and their three young children. With thunder and lightning booming and crackling and huge waves smashing across its bow, the captain ordered the passengers to get into the lifeboats as a precautionary measure. Dubois climbed in with his family but suddenly remembered that he had left a very important suitcase behind, a suitcase he believed contained the most important discovery to date of evidence of the evolution of the human species.

Leaving Anna and the panicking children in the boat, Dubois raced across the now empty deck of the steamer. Once inside his cabin, he grasped a large suitcase and brought it back on deck with him, lashing it by a strap over his shoulder. Climbing back into the lifeboat, he turned to Anna and said: 'If something happens to me, you must look after the children. I will look after this.'

The storm subsided and the lifeboats were never launched. Dubois climbed back onto the ship with the rest of the passengers, his precious suitcase hugged tightly to his chest. The declaration to his wife that a valise full of bones meant more to him than his own children did not exactly do wonders for his marriage, but it did not surprise her. Anna, more than most people, knew of Dubois's single-minded desire to show the world that he had found the missing link—the transitional creature that stood between the world of apes and the world of humans.

The Neanderthal debate

In 1856 two quarrymen dug up some very odd-looking bones—a skullcap, two femora, parts of a right and a left arm, and fragments of ribs—in a mine in the Neander Valley, in Germany. They assumed they were those of a bear but gave the fossils to a local naturalist, who in turn gave them to anatomist Hermann Schaaffenhausen. He understood that he was looking at, not a bear, but a human, perhaps the earliest human. (Actually, Neanderthal skulls had been discovered much earlier than this, in Belgium in 1829 and in Gibraltar in 1848, but in both cases they were not recognised as such until later.)

With the 1857 announcement of this discovery—and the subsequent discovery of two nearly complete Neanderthal skeletons at Spy, Belgium—the debate over the origins of the Neanderthals was on. The famous German pathologist Rudolf Virchow claimed that the Neanderthals were merely early human beings who suffered from rickets, hence the somewhat bowed legs and bent posture evident in the skeletal finds, but this ignores the fact that Neanderthal bones are perhaps fifty per cent thicker than those of modern humans, indicating a good deal more calcium than a sufferer from rickets would have in his or her body. (It is true that Neanderthals seemed to suffer an inordinate number of fractures—especially rib fractures—but modern researchers feel that this was because they may have leaped upon their prey with knives in order to bring it down—a hazardous method of hunting, at best.)

More Neanderthal finds were made in the twentieth century. We now have intact skeletons of children, a fossilised Neanderthal footstep captured for all time, tools that include spears and chisels, and Neanderthal hearth sites. More and more scientists have become convinced that Neanderthals, with their pronounced brow-ridge and heavy build (although they probably did not walk stooped, as previously thought) were only distantly related to Homo sapiens. *However, DNA shows about a 99.5 per cent genetic similarity between the two species. The current thinking is that Neanderthals and humans shared a common ancestor, but that the two diverged, with the Neanderthals emerging perhaps some 600,000 to 350,000 years ago.* Homo sapiens *then followed, perhaps 200,000 years ago.*

For a time the two species co-existed, especially in Europe, but then the Neanderthals became extinct, roughly 30,000 years ago. No one is quite sure why. Some scientists claim that they were killed off by humans; others say they simply interbred with Homo sapiens—*although current DNA studies show no signs of significant interbreeding. It may simply be that human beings were far more numerous and far more proficient at hunting and gathering, and so pushed the Neanderthals farther into remote areas where they lived and died in small groups.*

'The highest and most interesting problem'

In 1859 Charles Darwin published his seminal book *On the Origin of Species*, which put forth the theory of natural selection and evolution. But nowhere in the book did he touch on human evolution. In 1857 Alfred Russel Wallace had written to him, asking him whether he would discuss the origin of humans and Darwin had replied: 'I think I shall avoid the whole subject as surrounded with prejudices, although I fully admit it is the highest and most interesting problem for the naturalist.'

Although Darwin knew that human beings had hereditary variation in each generation, as well as varying numbers of children—two important ingredients of evolution—there was no extensive fossil record of human evolution, as there was of the evolution of plants and animals. In 1856 German miners working in the Neander Valley, near Düsseldorf, discovered among the bones of what turned out to be extinct cave bears and mammoths an extraordinarily thick human-seeming fossilised skullcap with a pronounced brow ridge. Other bones were found as well. This discovery and the debate about whether the so-called Neanderthal Man was a distant human ancestor or of another race entirely focused popular attention on the issue of the evolution of humans.

In 1871 Darwin did address human evolution head-on when he published *The Descent of Man*, in which he put forth the controversial thesis that humans had evolved from a shared ape ancestor, probably in Africa, and had been evolving ever since, especially through sexual selection, in which women chose fathers for their children on the basis of certain desirable traits, such as strength and hunting skills. The discovery of further and more extensive fossilised male and female Neanderthal skeletons at Spy, Belgium, in 1886—along with primitive tools—was more proof that an early race of human-like creatures had existed, perhaps hundreds of thousands of years before.

Closely watching the debate around this latest discovery as it raged through Europe—for then, as now, many people took the theory of evolution in regard to human beings as an affront to their religious beliefs or as unproven science—was a young Dutch anatomist and teacher named Eugène Dubois. Dubois, newly married, was a devotee of Charles Darwin's work, and he felt that the discovery of the Neanderthals showed that humans had indeed evolved over many millennia. But his ambitions lay even further in the distant past. Neanderthals were merely an early form of human being; what the ambitious and single-minded Dubois sought was the half-simian, half-human creature—the missing link between apes and humans—that would once and for all provide incontrovertible proof of evolution.

'What would be the most important discovery?'

Eugène Dubois was born in 1858 in Eijsden, a village in the southern part of The Netherlands. His father was a druggist and local politician; his mother was a member of a good local family that had lived in the village for generations. The Dubois family were respectable, well-to-do bourgeoisie (their family motto was *Recte et fortiter*, or 'Straight and true') and Eugène was raised with the expectation that he might follow in his father's footsteps. But Dubois, a smart, driven student with an aptitude for science, was fired by the current debates over evolution, which pitted those who believed that God had created the world and placed humans at the head of all his creatures against those evolutionists who claimed that humans had simply risen, through the often ignoble process of natural selection, above their fellow creatures. It seemed to the young Dubois that this was the most important debate that a scientist could have—one that reached right to the very heart and essence of existence. He wanted with all his heart to be a scientist who would make the discovery that would prove Darwin's theories once and for all. Even at the age of ten, he would sit down and write lists in which he asked himself the question: 'What would be the most important discovery?'

At the age of nineteen, in 1877, Dubois went to the University of Amsterdam to study medicine. Graduating at the top of his class, he accepted a teaching assistantship in anatomy at the university, working under a respected professor, Dr Max Fürbringer. It was not long before Dubois found out that life in academe was every bit as tough as life for an evolutionary animal in the jungle. When he wrote a brilliant paper suggesting that the cartilaginous structure of the larynx in mammals derived from gill cartilage in fish, Fürbringer insisted that he, too, be given credit for the idea. Dubois reluctantly agreed, but he bitterly resented his professor's unethical appropriation of his own ideas. Even though he was promoted to instructor in 1886—an important position for the newly wed Dubois, who had a child on the way—he understood that he would never be able to strike out on his own, without interference from academic tyrants like Fürbringer, unless he left The Netherlands.

The debate over the Neanderthal bones found in Spy, Belgium, focused his imagination on finding the missing link and gave him what his most recent biographer, Pat Shipman, calls 'a breathtakingly simple and bold idea'. Most fossilised finds of human or Neanderthal bones had simply been accidents. Now, Dubois decided to become the first person to deliberately set out to find 'the fossil record of human evolution'. In his typically logical way, Dubois methodically narrowed the world down to the place where he would search. In his opinion, Neanderthals were not primitive enough, not close enough to the apes to be the missing link. Therefore, he needed to search in a country where apes live or lived. He knew that the fossilised skeleton of an ape had been found in the Dutch East Indies. This skeleton, and fragments of others, was from the right period (the Pleistocene, when humans probably began to evolve into their present form). In this area—in Sumatra or Java—Dubois reasoned that he might have the best chance of finding the missing link, especially given the prevalence of limestone caves, which would have been used as shelter and tended to preserve skeletons better than open air sites.

'Old, old bones'

It was an extraordinary move for a young man of Dubois's background to make. Giving up his very promising academic career, he joined the army as a surgeon with the understanding that he would be posted to the Dutch East Indies. He and his wife, Anna, already had one child and she was pregnant with another—and, with the prevalence of tropical fevers and deadly diseases such as typhoid in the Indies, he was putting their lives at risk. And most experts in the field—even men friendly to Dubois—thought looking for a missing link was like searching for a needle in a haystack. Human or ape skeletons did not fossilise as well as animal skeletons did, even in limestone caves, and Dubois would merely be searching blindly.

Eugène Dubois

But nothing could stop the single-minded Dubois. He and Anna set sail for Sumatra in October of 1887, arriving in Padang, a Dutch colonial port on the Indian Ocean, in mid-December. Dubois was at first so busy treating illness at the military hospital that he found it hard to find time to make trips into the bush to look for fossils. However, he spread the word among the Malay village headmen that he was looking for 'caves with bones, old, old bones that are like rock. I want the bones of old animals that do not live here anymore.'

Frustrated with the abbreviated time allotted for his searching, Dubois published an article in a scientific journal summarising his reasons for believing that the missing link could be found in the East Indies. He ended by saying:

> *It is obvious that scholars from other countries will soon realize the promise of the East Indies. They will come and search for important fossils here and will find them, unless the Dutch authorities do something more to support such scientific work. And will the Netherlands, which has done so much for the natural sciences of the East Indian Colonies, remain indifferent when such important questions are concerned, while the road to their solution has been signposted?*

With typical determination, Dubois sent a copy of the article to the governor of Sumatra, who agreed to make forced native labour available and also facilitated the transfer of Dubois, Anna and their two children to a highland army station, Pajakombo, which was not quite so pestilential with fever. There, another doctor generously undertook a good deal of Dubois's workload at the hospital, and he finally had the time to search seriously for his missing link.

Into the tiger's lair

Beginning in August of 1888, Dubois took repeated expeditions of native labourers out with him to search the limestone caves of the hilly backcountry of western Sumatra. Typically, he taught his crew to find a cave—the limestone walls keeping it quite dark and cool, even in the torrid climate—and inside it dig a square hole with vertical walls, so that Dubois could see what strata his fossil finds lay in. As he suspected, the caves of Sumatra were rich with Pleistocene fossils, mainly the teeth of elephants, deer, tigers and buffalo.

His workmen did not, in the main, understand what this strange European was looking for, but they respected his courage. At one point, he found a long, narrow cave entrance on the side of a hill but was surprised when none of the labourers would go inside. With typical impulsiveness, he grabbed a lantern and shoved himself through, squirming on his stomach until the cave enlarged enough for him to stand up in it. He held the lantern high and realised by the rotting carrion and piles of gnawed bones that he was in a tiger's den—a modern tiger, not one from the Pleistocene. The tiger was not at home but might be at any moment. Controlling his panic, he squirmed back out of the cave, feet first in case the tiger should be trying to get in. As he neared the exit he discovered, to his horror, that he was stuck. Shouting as loud as he could, he finally was able to make his labourers understand that they should pull him out. When they did, he found out they had known all along he was entering the tiger's den, but since he had not asked them, they did not tell him.

Dubois had to face not only tigers but also malaria, which, despite his intake of quinine, would plague him during his entire stay in the East Indies. In December 1888, as he was recuperating from one such attack of fever, he received a letter from C. P. Sluiter, who was curator of the Royal East Indies Society of Natural Science in Batavia, Java. Accompanying the letter was an ancient skull. Sluiter explained that a Dutch mining engineer had found the skull near the Javan village of Wadjak. The engineer, not quite knowing what it was, had sent it to Sluiter and Sluiter was forwarding it to Dubois in the hope that he might be able to identify it.

Dubois immediately identified it as a human skull, probably from the early primordial people who had inhabited Java. The fact that such a skull could become fossilised and survive gave him hope for his own search, but he realised that he was probably searching in the wrong place. After some deliberation, he decided to apply to the Dutch government for permission to move himself and his family to the area around Wadjak, in the volcanic region of eastern Java. It was granted and he moved his wife and, now, three children to Java in April of 1890.

The Charnel House

At the time he moved to Java, Eugène Dubois was thirty-two years old, a blonde, well-built man with an air of physical strength about him, particularly in his upper body. Based on the animal fossils he had found already, the Dutch government had given him a three-year contract to allow him to devote his time to searching for fossils. Dubois realised that if he were ever going to find the missing link, the next three years were going to be the time to do it.

Slowed down during the latter half of 1890 by his recovery from malaria, Dubois entered the field again with a vengeance in 1891. He had decided to abandon cave digging for the time being. The skull sent to him by Sluiter had been found along a riverbank, and Dubois realised that the rising and falling waters of the slow Javan rivers might have preserved many a fossil in the sandstone riverbanks.

In the summer of 1891 he explored down the large and slow-moving Bengawan Solo River, searching for anything glinting white along the bank. Near the little village of Trinil, he paused to stare over a spit of land projecting from the bank into the river. Not quite knowing why, he decided to stop there to dig. His crew immediately began to find fossils of numerous species of animals; Dubois decided to call the place 'the Charnel House', since he was convinced many of these beasts were killed in one of the plentiful volcanic eruptions of the region and then preserved by the falling ash. And, in September of 1891, his men discovered something that made Dubois ecstatic: a tooth. Encased in stone, it was a third right molar from the upper jaw of an ape-like primate, but one that was not, quite, as old as one that had been discovered in India.

In October, while digging in the bank not far from where the molar had been discovered, Dubois's men found what at first appeared to be the top shell of a turtle, or perhaps half an ancient coconut. It was dark brown in colour, fossilised and encrusted with stone. But Dubois immediately recognised it for what it was: the top of a skull of some kind of higher primate. As he wrote soon after:

> *Near the place on the left bank of the river where the molar was found, a beautiful skull vault has been excavated ... As far as the species is concerned, the skull can be distinguished from the living chimpanzees: first because it is larger, second because of its higher vault ... It is clear from this what a gain to science the Javanese fossil is ...*

Although he could not yet prove it, Dubois was certain that he had found the fossilised skull of the missing link. The skull-top was too large, its cranial capacity too big, to be that of a chimpanzee, yet it was still too small to be human. The find had been made just as the rainy season came in, bringing work at Trinil to a halt, and so Dubois was forced to wait out the winter before returning there. In the meantime, he spent hours scraping away stone from the skull with various sharp instruments, careful not to shatter or damage the discovery that would, he was sure, shape his reputation.

'I have found it'

In May 1892, Dubois returned to work at Trinil, but while his crew uncovered fossil after fossil, they could not find another fossil from the unknown primate whose tooth and skullcap had been found in the previous years. Dubois came down with malaria yet again and had to return to his home. In August, while he was away from the dig, his labourers found a large, fossilised femur, which was sent to Dubois. He knew right away that it did not belong to an animal, but to a large, ape-like primate. The fact that it was found in the same geological stratum as the tooth and skull, and only a few metres away, indicated to Dubois that these bones came from the same creature.

Back at his home near Wadjak, Dubois measured the skullcap, finding that its cranial capacity was an estimated 1000 cubic centimetres, too large to be that of an ape. He also compared the skullcap with skulls belonging to a modern chimpanzee, to a human and to a gibbon. After carefully taking notes, he shut the bones up in carefully constructed boxes and told Anna: 'I have found it. These bones are the missing link. It is true, at last.' He then sent a report to the Dutch governor of the East Indies: 'I have, your Excellency, the honor of offering the first installment of the description of some of the fossils I have collected. This installment deals with only one species, *Pithecanthropus erectus*.'

Pithecanthropus erectus is the name Dubois gave to the missing link— it means 'upright ape-man'—and he decided that what he needed to do next was to bring his find back to the scientific community of Europe. More hardship lay ahead for Dubois and his wife in Java—an infant daughter was stillborn in 1893 and Dubois fought off even more bouts of malaria. Finally, in 1895, with his contract up, he was able to book passage back to Europe, braving the Indian Ocean storm with Java Man under his arm.

Fight for recognition

Back in The Netherlands, Dubois thought that his reputation would be made at last, but he found his discovery was greeted with a torrent of criticism from some, and with muted or conditional praise from others. Rudolf Virchow, the preeminent German pathologist, who weighed in so ponderously (and incorrectly) on Neanderthals, wrote that, in his opinion, the skull of Java Man was really that of a gibbon, while the femur belonged to a human being. As for Dubois's claim that the bones represented the missing link, 'Here the fantasy passes beyond all experience.'

In order to defend his find, Dubois travelled to London, Brussels, Liège and other European cities, carrying the bones in a suitcase and making speeches in front of scientists. (In Liège he left the suitcase behind accidentally in a cafe and experienced a moment of intense panic when he thought his life's work was lost; fortunately, a waiter had found the case and kept it safe for Dubois's return.) There were numerous scientists who believed his presentations, while others remained sceptical about whether the skull belonged to ape or human. (Dubois's response was to state that such confusion was natural, for any missing link would naturally have elements of both species.)

Part of the problem was that Dubois was not a member of the scientific/academic establishment at the time, and part of the problem was the prickliness of his personality. He was unable to charm or otherwise persuade people and he became angry when questioned too closely about his find. He was completely enraged when a supporter, a German scientist named Gustav Schwalbe, took a plaster cast of the skull (one Dubois had foolishly allowed him to make) on a lecture tour, in a sense completely appropriating Dubois's work and ideas.

So upset was Dubois at this and at the responses he was getting from the scientific community that he decided to stop showing the fossils completely. Around 1900, he put the bones away in a special, locked case at Teyler's Museum in the Dutch city of Haarlem (Dubois had become curator of palaeontology there) and concentrated on his work teaching at the University of Amsterdam and also measuring human cranial capacities. As other hominoid creatures, especially Peking Man, were discovered in the 1920s and 1930s, Dubois's identification of the skull as that of an ancient hominoid gradually gained favour, although modern scientists still are not sure that the skullcap, the femur and the molar are from the same individual, as Dubois claimed.

Eugène Dubois never quite recovered from risking his life to find *Pithecanthropus erectus,* only to find himself disdained at home, a victim of academic politics and professional jealously. He died in 1940, in a Netherlands occupied by the Nazis, with his place in scientific history still largely unacknowledged.

Pithecanthropus erectus and the missing link

Eugène Dubois's Pithecanthropus erectus, *or Java Man as his find came to be known, did not fare as well as Dubois had hoped when more and more knowledge about the descent of humans came to be understood by scientists, but in some ways the story is even more fascinating. In the decades since Dubois first found* Pithecanthropus erectus, *more and more fossilised bones of this type have been found, in Indonesia, in China—where the famous Peking Man was the most notable of the discoveries—and finally in Africa in the 1950s.*

However, scientists now no longer classify these skeletons as a missing link but rather as a separate species of Homo *known as* Homo erectus. *Recent, more sophisticated dating techniques put Java Man at about 700,000 years old, while some* Homo erectus *remains have been dated at an astonishing 2 million years old. But Java Man, or any member of* Homo erectus, *was not the missing link. That honour may go to a 47-million-year-old fossil of a lemur-like creature scientists have nicknamed Ida. It was discovered outside Frankfurt am Main before 1983, in a pit where numerous Eocene era animals had been preserved after a volcanic eruption.*

Scientists did not at first realise what they had found, but close study of the fossil indicates that it may be the missing link between apes and humans. Ida, about the size of a cat, has opposable thumbs and fingernails instead of claws, and has hind legs that indicate the ability to stand upright. Palaeontologist Jørn Hurum, who led the team that researched Ida, has said that she is 'the link to all humans. This is the closest thing we can get to a direct ancestor.'

MARY KINGSLEY
'Why has this man not been buried?'

Although Mary Kingsley died well over a century ago, her voice, coming through her classic works *Travels in West Africa* and *West African Studies*, is strikingly familiar to our ears. This is partly because she inspired an entire school of travel writers, one that persisted well into the twentieth century—the innocent-taking-on-grave-dangers-with-insouciance school—but also because she was so genuinely an original that her spirit cannot be diminished by the passage of a mere century.

Kingsley tells the story that when she decided to travel in West Africa her genteel acquaintances in late Victorian England were at a loss as to how to advise her. However: 'One lady remembered a case of a gentleman who had resided some years at Fernando Po, but when he returned an aged wreck at forty he shook so violently with ague as to dislodge a chandelier, thereby destroying a valuable tea service and flattening the silver tea-pot in its midst.'

It was for just this reason—in order to render chaos unto the stuffy tea service of her life—that Mary Kingsley, at the age of thirty-one, set out on her most unprecedented adventure, wearing the garb of her class (long, trailing skirts and high-necked blouses) but carrying few of its preconceptions. Her talents as a naturalist would lead her to discover seven species of fish unknown to science, while her love of Africa made her one of the few Europeans of her time who did not try to impose her own consciousness on the people of that continent.

'Life in the Torrid Zone'

Mary Henrietta Kingsley was born in Islington, London, in 1862, into an interesting but highly dysfunctional family. Her father, George Henry Kingsley, was a doctor who spent the better part of his life travelling the world in the company of various members of the English nobility. He was fascinated by the sacrificial rites of native peoples and loved the tropics. Mary was to write: 'The sunlight, the colour, and the magnificent exuberance of the life in the Torrid Zone absolutely called across the latitudes to every member of the Kingsley family ...' It was George Kingsley's habit to spend only a few months of the year, at most, back in England; during one of these periods, in 1862, he rather bafflingly married thirty-five-year-old Mary Bailey, his cook. Baffling to young Mary Kingsley, that is, until she discovered, after the deaths of both of her parents, her birth certificate, which indicated that she had been born four days after her parents' wedding.

Three weeks after the marriage, George Kingsley set off wandering again, returning home for just a few weeks each year (during one of these sojourns siring Mary's brother, Charles). His financial support of the family was erratic and his wife was not readily accepted by the Kingsleys; eventually, she took to her bed with a variety of illnesses more imagined than real, shutting the real world out and leaving her daughter to play nursemaid to her in the best Victorian fashion.

This Mary did. And when her father refused to send her to secondary school—although he later spent two thousand pounds to educate her brother at Cambridge—she spent all her spare time in the library, reading voraciously, taking a particular interest in natural history, medicine and science. In 1886 George moved the family to Cambridge, in part so that Charles could attend Christ's College, but also because his health was failing and he wanted his daughter to put together his huge collection of notes on fetishes and tribal sacrificial rites. This Mary dutifully attempted to do, but her father died in 1892 with the task still incomplete, and when her mother died just six weeks later, Mary found herself finally free—or relatively so, since she still had (by Victorian custom) to take care of her brother. But in 1893, when young Charles decided to travel to China, Mary was now 'in possession of five or six months which were not heavily forestalled, and feeling like a boy with a new half-crown, I lay about in my mind ... as to what to do with them'.

A true Kingsley, she decided the answer lay in the tropics—in Africa, to be exact.

The fabulous Kingsleys

Mary's father, George Kingsley, came from a family with a chequered and sometimes tragic history. He had four brothers, Charles, Herbert, Gerald and Henry. Herbert disappeared early from the scene—at the age of fourteen he was sent to boarding school where he stole a silver serving spoon, sold it and was caught. Disgraced, he suffered the Dickensian fate of being locked in a room and fed only bread and water. Within two months he died, supposedly of rheumatic fever, although rumours persisted that he had actually escaped from the room and drowned himself.

Next went Gerald, who was an officer in the Royal Navy, sailing aboard HMS Royalist, *a gunboat sent to patrol the Torres Strait, off the coast of Australia, in 1844. The ship was becalmed and stranded in the Gulf of Carpentaria for a horrible year and a half, with the sailors resorting to eating their own dead to stay alive. Gerald was the last of the ship's officers to die and was very probably eaten.*

Charles and Henry both became writers, Charles being the most famous, well known in the nineteenth century as the author of the children's book

The Water Babies, *although his literary output is mainly forgotten today. While a religious man (as well as being a writer, he was an Anglican minister who was chaplain to Queen Victoria), Charles was very probably bisexual and a man with some very strange sexual tastes—he practised self-flagellation and once sent his wife a drawing of the two of them having sex on a crucifix while they ascended to heaven. He was manifestly racist—he referred to both the Africans and the Irish as 'human chimpanzees'—and died in 1875 covered with honours.*

Henry was also a writer, although a less talented one than Charles. He went off to Australia in 1853 to strike it rich in the gold rush but did not have much luck. Although he did write a book called The Recollections of Geoffrey Hamlyn *(1859), which made him briefly famous, he soon settled into a career of hack work that left him quasi-penniless. He died of throat cancer at the age of forty-six after a lifetime of heavy drinking and smoking.*

'I went down to West Africa to die'

Why Africa? There were a number of reasons at work. Her stated and rather facetious rationale for the journey was the fact that Malaya was too far and South America full of disease, but since West Africa was no slouch in the disease department—typhoid, malaria, beriberi and any number of other illnesses were just waiting to snare travellers—this cannot be believed. Kingsley did record in her journal that she wanted 'something to do that my father had cared for', which may be closer to the truth. Kingsley herself was fascinated by African fetishes—objects like bones, fur, statues, feathers, dolls and diamonds that are thought, especially in West Africa, to have supernatural powers. Too, she had been raised, like many British people, on the widely circulated stories of African explorers such as John Hanning Speke, Richard Burton, Henry Morton Stanley and the great Dr David Livingstone.

Kingsley also took a brief trip to the Canary Islands, off the northwest coast of Africa, just after her parents died. One reason she went there was that it was a regular stopping place for visitors to and from West Africa—missionaries, traders, government officials. She joked that all their stories of people they knew in the region invariably ended with the line 'but he is dead now', although she could see that most of them were passionately in love with Africa.

Finally, there was a darker reason. In 1899, looking back on her travels, she wrote to a friend named Matthew Nathan—a British officer who eventually became governor of Sierra Leone—that after her parents died and her brother decamped to China 'she was dead tired and feeling no one had need of me anymore ... [so] I went down to West Africa to die. West Africa amused me and was kind to me and was scientifically interesting—and did not want to kill me just then ...'

This contains more than a touch of melodrama, but it is possible that Kingsley may have been heading to Africa to see if she could survive its rigours—and if not, she would join her parents in the grave. Whatever her reasons, when she decided to really make the journey, she put her all into it. She immediately began canvassing her acquaintances for information, all of which, she wrote, came under these six headings:

> *The dangers of West Africa*
> *The disagreeables of West Africa*
> *The diseases of West Africa*
> *The things you must take to West Africa*
> *The things you find most handy in West Africa*
> *The worst possible things you can do in West Africa*

Looking for a guide to the language, Kingsley found 'a French book of phrases which contained the following "common" sentences needed by intrepid travellers to the region: "Help, I am drowning.";
"Get up, you lazy scamp!"; "Why has this man not been buried?"'
(The answer, according to the book, would usually be: 'It is fetish that

has killed him, and he must be exposed with nothing on him until only the bones remain.') Not surprisingly, she decided not to take the book with her.

More practical advice came from an old African hand, who told Kingsley:

> *When you have made up your mind to go to West Africa the very best thing you can do is get it unmade again, and go to Scotland instead; but if your intelligence is not so strong enough to do so, abstain from exposing yourself to the direct rays of the sun, [and] take four grams of quinine every day for a fortnight before you reach the Oil Rivers [Nigeria] ...*

'A hundred other indescribabilia'

On 31 July 1893, having made out her will, Mary Kingsley sailed for Africa aboard a cargo steamer, the *Lagos*. She took with her a black medical kit, a portmanteau, a large waterproof duffel bag, a bowie knife and a revolver. At the Canary Islands, the only two other women on board—one the wife of a British officer, the other a stewardess aboard the steamer—left the ship and Kingsley travelled on as the only female aboard. Fortunately, the captain of the *Lagos*, John Murray, took her under his wing, brought her to dine at his table, and became something of a mentor to her, seeing, as she wrote, 'that my mind was full of errors that must be eradicated if I was going to deal with the Coast successfully'.

The steamer arrived in West Africa in August, during the rainy season, when the sky was 'wild and strange' with fierce, towering thunderclouds. They anchored off the coast of Sierra Leone in the middle of a thick mist, with the sounds coming from the shore tantalising Kingsley. The next morning, she went ashore in Freetown, a collection of dilapidated but colourful shacks that housed the descendants of the freed slaves from British Canada, Britain and the West Indies. Immediately, she was caught up in the swirl of West

African life. It was market day, raucous and chaotic, with chickens, fish, fruits and vegetables and 'a hundred other indescribabilia' for sale. The place was full of marauding crows 'who gave the impression of having been extremely drunk the previous evening', and goats and sheep wandered everywhere, along with 'small, lean, lank yellow dogs with very erect ears'.

Kingsley was accompanied by Captain Murray, who was taking her to meet a British trading agent, for she had the unusual idea—one should say unusual for a woman at the time—to make her way into the interior as a trader. She had brought with her glass beads, fishhooks, tobacco and the like to exchange for food and to pay porters. The British trader in Freetown was supposed to advise her about this, but first he and Captain Murray sat and drank iced whiskey together in the shade in front of his store, during which time the captain was pecked by an unruly ostrich and Kingsley, startled by a semi-feral monkey, fell into a cellar, fortunately landing on bales of cotton goods. Shortly thereafter, to make her welcome complete, a plague of locusts hit the town.

'Reflections in a mirror'

All this occurred in a single day, and that evening the *Lagos* left Freetown, heading south along Africa's fabled Ivory and Gold Coasts, making stops in Liberia—another haven for freed slaves, from America—Accra and Benin. The voyage was Kingsley's initiation into Africa. It poured rain ('everything on board was reeking wet,' she wrote, 'and the towels too damp on their own account to dry you'), but as the coastline slipped by, she became fascinated by the monotonous yet deeply moving vista, which was composed of four different lines, or 'bands', when seen from a distance, 'lines that go away into eternity as far as the eye can see':

> *There is the band of yellow sand on which your little factory [trading post] is built. This band is walled to landwards by a wall of*

dark forest, mounted against the sky to seaward by a wall of white surf; beyond that there is the horizon-bounded ocean ... In the light of the brightest noon the forest-wall stands dark against the dull blue sky, in the depth of the darkest night you can see it stand darker still, against the stars ... Night and day and season changes pass over these things, like reflections in a mirror ...

Along the coast, Captain Murray stopped to pick up a large number of passengers who were mainly Kru people, who lived near the sea and often worked for the British. Naked, they paddled canoes out of hidden inlets when they saw the *Lagos* and made straight to the ship, singing and chanting, and then swarmed aboard by climbing up the sides. Deciding it might be better to retreat to her cabin for a time, Kingsley discovered it was already full of Kru, who were drying themselves with her towels and admiring her hairbrushes. She finally got them to leave, but the new passengers made so much noise that she could not sleep that night and so decided to go on deck and fish—fishing had always been an occupation that calmed her nerves. However, in this case, she managed to hook a 2.5-metre (8-foot) octopus, which climbed up the side of the ship and appeared to be intent on chasing her until, with the help of a fellow British passenger, she beat it back into the ocean with paddles.

The *Lagos* now reached the Oil Rivers—so named for the palm oil trade on them—that led into the interior of what is now Nigeria. It stopped briefly at a trading post, where the Europeans kept themselves as drunk as possible in order to ward off an epidemic of yellow fever (such measures did not work, as nine out of eleven white men died there within ten days). Glad to be gone, the *Lagos* sailed farther south to Loango in the French Congo, where Kingsley intended to hire porters and strike off into the interior on her own. However, intimidated by the darkness and violence of the place—the only place she could find to stay was with the British consul, who was also putting up a horde of drunken sailors from a wrecked ship—she hopped the nearest freighter she could find and made her way to the town of Cabinda, in Angola. There she met an English trader and writer named R. E. Dennet, who, as had Captain Murray, took her under his wing and helped tutor her in the mysteries of Africa, particularly in West African religions.

Word spread of the arrival of a lone white woman who knew something of medicine—Kingsley carried with her a medical kit, and her years of nursing both parents had provided her with a great deal of experience. One night a native healer, or 'fetish priest', made his way in from a neighbouring village and begged her to return through the forest with him to treat a desperately ill patient. She agreed, even though, as she said, the woods 'were an alarming place to walk about at night, both for a witch doctor who believes in all his forest devils and a lady who believes in all the local, material ones'. But Kingsley persevered in this powerful initiation into Africa, found that the sick man had a disease (she does not identify it) that could be easily treated by the drugs she had with her and, pointedly, gave the healer the drugs to administer.

The man walked her back through the forest to the house she was staying in and, as they walked, they conversed about his fetish objects and his animistic faith. Kingsley felt that none of this was foreign to her, as it might be to another European person—that, in fact, she identified with the man's gentleness and his profound love of nature. It was here that she began to realise that she 'was more comfortable [in Africa] than in England'.

After further adventures she returned home to London in early January of 1894, but she realised quickly that her heart belonged now to Africa.

Return to Africa

Kingsley returned to the small flat she now shared with her brother, Charles, who was back from Asia, but found that she could not adjust to life in England again—the cold, damp weather, the stuffy gentility, the sense that life lacked all excitement and drive. She busied herself by filling her rooms with souvenirs, so much so that the place turned into a virtual shrine to Africa, and by writing a story of her travels that George Macmillan, head of Macmillan Publishing Ltd (who also happened to be her uncle) desperately wanted to publish. At first she agreed, but she wanted to put the book out only under a pseudonym or just using her initials, so that no one would know her sex—she rightly envisioned Macmillan using her 'trail of petticoats' over Africa as a way to promote the book as a curiosity.

And Mary Kingsley had more serious intentions than that. She was a naturalist whose specimens of fish and insects were happily accepted by the British Museum and who was, in fact, now commissioned to find fish in the area of the Ogooué River by none other than the museum's head of zoology, Dr Albert Charles Günther. Therefore, she forbade George Macmillan to publish—yet—and returned to Africa in late January of 1895 aboard the *Batanga*, once again skippered by her good friend Captain Murray. Although, backed by her publisher, she had more money now and was travelling, at the beginning of the trip, with her friend Lady Macdonald, she still intended to enter the interior of Africa as a trader, swapping tobacco and beads for ivory and using the ivory to pay for food supplies and porters.

After Macdonald joined her husband, Mary Kingsley once again travelled by steamer south down the coast of Africa. She was heading to the French Congo, where she intended to travel up the Ogooué River to collect specimens of fish 'from a river north of the Congo'— in other words, to find new fish to preserve and present to Günther at the British Museum. She travelled by steamship up the Ogooué as far as the trading post at Lambarene. From there, her West African adventures truly began, for now she hired eight Igalwa natives to paddle her farther up through heavy rapids—this despite a warning from local French people that 'the blood of half my crew is half alcohol'. She was travelling into the land of the cannibalistic Fang people and sensed that the Europeans she was leaving thought she might not return: 'On the whole, it is patent that they don't expect to see me again, and I forgive them, because they don't seem cheerful over it.'

Kingsley was, after all, the first white woman to travel alone to these regions, and she would be the only white person some of the indigenous people she met had seen. Yet she seems to have had an innate sense of how to deal with Africans and how to live among them. She always ate the local food—usually manioc meal and smoked meat that had 'the toothsome taste and texture of a piece of old tarpaulin'. It was not that she was ignorant of the dangers she faced, however. She later wrote of her time in the African wild:

> *You walk along a narrow line of security with gulfs of murder looming on each side, and where in exactly the same way you are as safe as if you were in your easy chair at home, as long as you get sufficient holding ground; not on the rock in the bush village, inhabited by murderous cannibals, but on ideas in those men's and women's minds; and these ideas, which I think I may say you will always find, give you safety.*

This is an extraordinary statement for an English person of Kingsley's time, to wit, that if you treat those you encounter like human beings, they will generally treat you in exactly the same way.

'A little something ... as a memento'

By the end of July, travelling by canoe and foot, Kingsley had reached the land of the Fang, near Lake Ncovi, although she was suffering from a bout of malaria. She discovered them to be, quite literally, a noble people: handsome, intelligent, brave and kind. Many of them had never before seen a white person, man or woman; the children, especially, threw their hands up in horror and dismay when Kingsley approached. Travelling through with guides, she saw numerous elephants and other game, including an 'impressive [snake], about three feet six inches [1 metre] long and thick as a mans thigh'. A Fang warrior smashed the snake's head with one thrust of his gun butt and stuffed it overflowing into his pouch. Kingsley ate it for supper. One day she was walking along a trail when she found herself falling like Alice in Wonderland, except it was not a rabbit burrow she tumbled down but a camouflaged 4.5-metre (15-foot) deep game trap, whose bottom was lined with spikes, each about 30 centimetres (12 inches) long. Fortunately, her Victorian garb of thick skirts and blouse protected her, and she could not help but crow: 'It is at these times you realize the blessing of a good thick skirt. Had I paid heed to the advice of many people back in England who ought to have known better ... and adopted masculine garments, I should have been spiked to the bone and done for.'

Kingsley's Fang guides hauled her out with a rope and, none the worse for wear, she went on her way. A little bit more unsettling was the night she spent alone in a Fang hut kindly provided by her hosts. After a short time, she became aware of a stench of 'an unmistakable organic origin'. Looking around her, she realised it was coming from skin bags hanging from the ceiling. Using a lighted torch to investigate, she dumped out the contents of the biggest bag and found 'a human hand, three big toes, four eyes, two ears, and other portions of the human frame. The hand was fresh, the others only so-so, and shrivelled.'

Carefully putting the contents back in the bag and tying it up exactly as she found it, she reflected that 'although the Fan [Fang] will eat their fellow friendly tribesfolk, yet they like to keep a little something belonging to them as a memento'.

Ultimately, as Kingsley left the land of the Fang to make the long journey back towards civilisation, having crossed overland to the Rembwe River, she thought: 'A certain sort of friendship [had arisen] between the Fans and me. We each recognized that we belonged to that same section of the human race with whom it's better to drink than to fight.'

'It crouched down to spring'

Mary Kingsley came across a good deal of dangerous wildlife during her time in West Africa and, unlike some unfortunate visitors, lived to tell the tale, although sometimes just barely. She survived driver ants, vast invading columns of which would take over African villages, entering hut after hut to devour vermin—and anything else—in their paths. She fought off huge crocodiles with just a paddle for defence. But one of her closest calls with dangerous animals was during her first trip to Africa, when she had agreed to help some Sisters of Mercy nurse sick traders back to health in the settlement of Bimbi, near Rio del Rey, just before she returned home to England.

While helping in the settlement, Kingsley often took the night shift, sitting up among the delirious and moaning men. The men were sheltering in a kind of open ward in a house built on sticks on the banks of a river in the jungle. The house belonged to the head trader of the area, a German. One night, she heard the man's boarhound barking and growling. She walked out on the verandah to find a huge black leopard with its teeth sunk into the dog.

Many in Kingsley's position would have turned and run to a safe distance from the leopard, or at least grabbed the nearest gun, but she was made of different stuff:

> I went to the rescue with a chair, which I let into the leopard so lustily that the intruder let go its hold on the dog and turned on me ... It crouched, I think, to spring on me. I can see its great, beautiful, lambent eyes still, and I seized an earthen watercooler and flung it straight at them. It was a noble shot; It burst on the leopard's head like a shell and the leopard went for bush one time.

She measured the leopard's paw prints the next morning in the soft dirt surrounding the house and realised that the animal had been 3 metres (10 feet) in length.

The Throne of Thunder

After heading down the Rembwe to the coast and the capital city of Libreville—taken in a sailing canoe by a colourful African trader named Obanjo, who preferred to be called Captain Johnson— Kingsley headed up the coast to the German colony of Cameroon where, in September, she decided, more or less on a whim, to climb Mount Cameroon, which is 4100 metres (13,450 feet) high and was known to the Africans as *Mungo Mah Lobeh*, or the Throne of Thunder. Ascending through a forest filled with the buzzing of thousands of bees, she and her guides finally reached 'the great south-east wall' of the mountain and she and a few of them began to climb through the clouds into the blazing sun. Kingsley outstripped those with her, until they began to look like 'little dolls' far below her, and at the top of the wall found 'a great lane, as neatly walled with rock as if it had been made with human hands'. This led her to a strange, rock-strewn plain from which she could see 'the South Atlantic down below, like a plain of frosted silver'.

The view was incredible but, with the drama that usually attended Kingsley's life, it was soon hidden by a ferocious thunderstorm that drove her partway down the mountain to safety. However, she spent the next two days doggedly climbing upwards—at one point spotting a tornado in the distance—and finally made it to the windswept peak. There she found a clutter of champagne bottles left behind by others who had climbed the Throne of Thunder; to this she added her calling card, and then she turned and made her way back down the mountain.

Shortly thereafter, Mary Kingsley sailed for England, arriving back home in November with a pet monkey on a leash. To her great surprise, a journalist was there to meet her, the first of many. Stories of her journeys—her 'trail of petticoats'—had been circulating in England (in part through letters she had written to friends, who had released them to newspapers) and she realised she was famous. She cared almost nothing for fame, but she did need money, both to help support her and her brother and also, most importantly, to return to Africa.

Kingsley now began work in earnest on the book that became *Travels in West Africa*, which was published by Macmillan in 1896 and immediately sold out several editions. She started to do a series of immensely popular lectures on Africa (the first one was given to the staff of a London medical school and was entitled 'African Therapeutics from a Witch Doctor's Point of View'). Yet her popularity, her sex and her droll persona obscured her very real accomplishments. On both journeys through Africa, Kingsley had gathered over a hundred species of fauna, including seven species of fish unknown to science. She used her newfound fame to become a powerful voice on behalf of Africa and Africans, voicing numerous opinions that went against conventional wisdom in Victorian Britain. Missionaries, she said, merely tried to shape Africans in their own image, rather than appreciating them for who they were—it was the African traders who were best at bringing civilisation to the country. She even defended polygamy as a necessary practice in a country where women nursed children until they were two years old and would not have sex with their men during this time. In all this, her main concern was that African customs not be destroyed by encroaching 'civilization'.

Mary Kingsley could not stay away from Africa long, however. She returned in 1899, just as her second book, *West African Studies*, was being prepared for publication. This time she headed to South Africa to collect fish specimens from the Orange River but found herself embroiled in the Boer War, which had just broken out. Typically, she turned to helping other people—in this case Boer prisoners-of-war who were suffering from typhoid fever. Unfortunately, she came down with typhoid herself and quickly became severely ill. With typical modesty, she asked only for two last favours from the doctor and nurse who were taking care of her. One was to allow her to die alone—by herself, for, as she told them, it was a journey she needed to make alone, like so many of the journeys of her life.

The second favour was to be buried at sea, rather than back in England. After her death on 3 June 1900 at the age of thirty-seven, her body was consigned to the waters off the Cape of Good Hope.

Brycinus kingsleyae

While Mary Kingsley could converse learnedly on insects and loved to observe animals, fish were really her thing. After her second trip to West Africa, she brought back sixty-five species of fish, of which eighteen were new to Gabon (the area where she travelled on the Ogooué) and seven completely unknown to science. Three of these latter were named after her—Brycinus kingsleyae, Brienomyrus kingsleyae *and* Ctenopman kingsleyae— *by Dr Albert Charles Günther, the zoology head at the British Museum, for whom she collected.*

In the April 1896 issue of Annals and Magazine of Natural History, *Günther wrote appreciatively of Kingsley's quest, pointing out that before she arrived hardly anything was known of the species of fish to be found in the Ogooué. Kingsley herself downplayed her collecting of specimens, preferring to present herself as a relaxed angler simply passing some leisure time. In the chapter in* West African Studies *entitled 'Fishing in West Africa', she describes herself as a 'born poacher', who loved casting a line or scooping a net into the obscure lakes and rivers she came to, with the 'usual result' being fifteen baskets of fish in a very short time. Most of these were 'common mud fish', which made very good dinners; she collected her rare specimens deeper into the interior, by means of a trap she called 'the stockade trap'. This was done by driving stakes close together into the water and leaving one opening, upstream. The fish would swim into this, driven by the fast current, and then be unable to get out again. Kingsley occasionally had to fight crocodiles for the contents of her trap, but she always claimed that it was worth it.*

GEORGE FORREST
'of seed such an abundance'

There is a photograph of George Forrest taken deep in the mountains on the Chinese–Tibetan border, probably not long after his infamous 1905 expedition and his extraordinary escape from death at the hands of lamas bearing poison-tipped arrows. If you did not know who Forrest was, you might think to yourself: Aha, yet another Victorian big game hunter, kneeling there in his khakis, rifle and hunting dog next to him. Forrest's gaze is direct, almost pugnacious, his body stocky and muscular—his calf muscles positively bulge—and he looks ready to kill almost anything that comes into view. But—what is that sitting daintily in his hat? Oh, a daisy.

George Forrest was a hunter all right—a plant hunter, and one of the most famous of his time. The area he hunted was a wild no man's land in western China carved up by three great rivers—the Mekong, Yangtze and Salween—and the chasms and valleys down which they tumbled from their sources in the Tibetan highlands. Forrest was a self-taught naturalist who was one of the last plant hunters to work for a private collector, the famous Arthur Kilpin

Bulley, and who was, to the delight of rhododendron lovers everywhere, a 'rhodie man' through and through. Despite the fact that he worked in China during some of the most turbulent years in its history—'living in China right now', he wrote to a friend in 1912, 'is like camping alongside an active volcano'—he stayed there some twenty-eight years, always protesting to his wife back in Scotland that the next expedition would be his last. He took one more 'last' expedition in 1932 and never came back from it, which is probably the way he would have wanted it.

Rhododendromania

Jane Brown, the author of Tale of the Rose Tree: Ravishing Rhododendrons and Their Journey around the World, *has written that rhododendrons are 'too complex a genus for simple truths'. There are 1025 species that we know of, and they range in size from* Rhododendron caespitosum, *which was only a few centimetres long, to* Rhododendron magnificum, *found by Frank Kingdon-Ward in the Burmese foothills— its trunk was a metre in circumference and it featured a thousand flowers. The rugged 'rhodies', as aficionados love to call them, can be grown throughout most of the Northern Hemisphere and in many parts of the Southern Hemisphere, from high in the Himalayas to the lowlands of the American South.*

For many centuries travellers' tales of these 'voluptuous and perfumed flowers' piqued the curiosity of plant lovers, especially in Britain. Finally, in the mid-nineteenth century, Sir Joseph Dalton Hooker (see page 198) and Dr Thomas Thompson helped introduce the seeds of the rhododendrons of Sikkim to Britain, where they were grown successfully in greenhouses, and what has been called 'Rhododendromania' kicked off, with plant lovers clamouring for the new species. The craze continued into the twentieth century, when George Forrest was hired by the Rhododendron Society, which sponsored him on his last three trips. He was to end up introducing over three hundred new species of rhododendrons, enough to satisfy any rhododendromaniac.

A budding young botanist

George Forrest was born in Falkirk, Scotland, in 1873, the youngest son (out of thirteen children) of George and Mary Forrest. The elder George Forrest was a draper who sold bolts of cloth from a small shop, but in the late 1880s both his health and his business began to fail and he moved with his family to live with his eldest son, a Protestant minister, in the more rural town of Kilmarnock, in southwestern Scotland. This is where young George completed his education and began wandering through the countryside, developing his love of plants and flowers.

Forrest trained to be a pharmaceutical chemist, training that included learning the medicinal properties of plants. He was also schooled in basic medical and surgical procedures, which would stand him in good stead in the wilds of China. He did this work for a few years, but after his father died and he came into a small inheritance, he decided it was time to see the world.

His first stop was Australia, where he landed in 1891, at the age of eighteen. He was just in time for a gold rush in the western part of the country and tried his hand at prospecting, apparently with a modicum of success. But Forrest was too restless to stay in one place for long. He spent several years working on various sheep-stations, which is where he came to enjoy hard work and long hours in the open country. Then, after a stop in South Africa, he returned to Scotland in 1902.

Forrest did not keep a journal beyond a botanising field book and he died before he could write the memoirs that he was always claiming he would pen in retirement, and so much of what we know about him, especially at this stage, comes from letters to and from family and friends. While in Australia and South Africa, his youthful fascination with plants had returned and he had built up a fairly sizeable collection of specimens by the time he returned to Great Britain. He apparently decided that he wanted to combine training with employment, and thus applied for and got a job at the Royal Botanic Garden in Edinburgh.

Forrest worked in the herbarium, helping to dry and preserve plant specimens; he spent the next two years increasing his knowledge about flowers and impressing his superiors, who included the prominent botanist Sir Isaac Bayley Balfour. Balfour and the others who worked with Forrest were quite fond of the young man, who, in keeping with other botanists, had certain eccentricities. A desk job really was not for him, a problem he solved by walking to and from work each day, a round trip distance of almost 20 kilometres (12 miles). He refused to sit at his specimen table, standing to do all his work. And, as he did all his life, he wore knickerbockers instead of trousers.

'The best international collection of dandelions'

Forrest was finally able to break out of the bondage of his dreary desk job in 1904. Sir Isaac Balfour was approached by Arthur Kilpin Bulley, the wealthy Liverpool cotton merchant who had built a grand house near Neston, on the estuary of the River Dee and who was in the process of putting together the magnificent garden that is now the Ness Gardens of the University of Liverpool. He deeply wanted exotic plants and seeds from China, as he wrote to Balfour, but his attempts to get seeds informally from missionaries there had resulted in 'the best international collection of dandelions to be seen anywhere'.

Therefore, Bulley asked Balfour to recommend a collector to travel to China for him to bring back flowers, plants and seed, and Balfour suggested Forrest. Forrest had never actually worked as a plant hunter, but the combination of his outdoor experience in Australia and the knowledge gained working at the Botanic Garden proved just the right one. After arranging things with Bulley—who was one of the last of the private patrons for plant hunters, since most were financed by large corporate nurseries—Forrest sailed to Rangoon, in Burma, went by rail to Mandalay and then crossed into Yunnan Province, in southwest China, which was to be his main base of operations for many years to come. He then travelled to the town of Tengchong, arriving in August of 1904.

'All to Gather—All Together'

Arthur Kilpin Bulley, patron to George Forrest and Frank Kingdon-Ward, has also been called 'the great patron of 20th century plant introductions'. He was an interesting man. Born in 1861 near Liverpool, he was one of fourteen children of a wealthy cotton broker. He joined the family business at an early age and did quite well, but he had always loved plants and, when he went abroad, spent a good deal of time collecting them.

In 1898 he brought a 24-hectare (60-acre) tract of land at Ness and established a large garden. Despite his wealth, Bulley was a socialist and a supporter of women's suffrage; he opened part of his garden to the public as a kind of gardening cooperative, under the slogan All to Gather—All Together. In 1903 he founded Bees Nursery at Ness, from which he would sell plants he had raised from the seeds that were sent to him by the likes of George Forrest. By 1911 he had founded the famous Bees Seeds, making cheap packets of seeds available to the public. He died in 1942, but his daughter bequeathed his beloved gardens to the University of Liverpool; they are now known as the Ness Botanical Gardens.

There was a reason Forrest was going to this particular area of China. Watered by the Salween, Yangtze and Mekong Rivers flowing out of Tibet, the area features deep valleys with precipitous cliffs, where plants are isolated from other plants and thus grow into unusual and distinctive species. Because this part of Yunnan encompasses valleys, alpine meadows and mountain peaks, there is a wide variety of flora to be found there. Forrest was by no means the first plant hunter there—Ernest 'Chinese' Wilson and Augustine Henry had been there before him, and the French missionary priests attempting to work in the Tibetan highlands were also, in many cases, plant hunters—but he would become arguably the most successful.

It was too late in the season to do much collecting, but George Litton, the British consul in Tengchong, befriended Forrest and the two men went on a trip to a French missionary station in the Tibetan border country, where Forrest saw plant specimens collected by the priests and decided that he wanted to return there the following summer.

'A lot of killing'

In the early part of the twentieth century the wilderness areas of Yunnan Province were not necessarily the safest places in the world for a plant hunter, or any foreigner, to operate—Augustine Henry had warned that it would take 'a lot of killing' before enough bandits could be cleared out of the country to make it even remotely peaceful for travellers.

The year 1905 was a difficult one along the Chinese–Tibetan border for political reasons. In 1903 Colonel Francis Younghusband had led a British invading force into Tibet with the intention of establishing British control over the country, since Great Britain feared a Russian attempt to take over the mountainous kingdom. In 1904 that expedition ended disastrously with the British massacre of hundreds of Tibetan militia and Younghusband's unauthorised entry into the holy city of Lhasa. At around the same time, the Chinese took over the town of Batang. In the spring of 1905, the lamas in Batang rose up and murdered the Chinese officials stationed there. Stirring up the population, which consisted of numerous tribes of indigenous Tibetans, the lamas led a revolt with French missionaries as a particular target. Father Jean Soulie, a noted botanist, was captured in Batang in April, tied to a tree, tortured and shot. A short time later, in the nearby town of Atuntze, more missionaries were murdered and the Chinese troops there to protect them were wiped out.

For the moment unaware of this, George Forrest was deep in his first plant-hunting trip just two and a half days' walk to the south. He had stationed himself at the French mission in the little town of

Tzekou, on the bank of the Mekong River, at an altitude of about 1500 metres (5000 feet). He was a guest of Father Dubernard, the chief French missionary there, and two other elderly French priests. Forrest had spent the winter of 1905 training seventeen Chinese botanists and assistants in the art of locating plants and collecting and drying them, and he now gloried in roaming the upland valleys.

However, rumours of the violence to the north soon spread to Tzekou and on 19 July it was decided that everyone in the mission—this included the three priests, Tibetan converts, and Forrest and his seventeen staff members—should evacuate south to the town of Yetche, 50 kilometres (30 miles) away, where there was a sizeable Chinese garrison. They left in the middle of the night, following a precipitous track that took them near a lamasery. Unfortunately, someone in the missionary party made a noise, the alarm was sounded and the group was forced to run desperately through the night. The next morning they saw black smoke rising behind them and realised that the Tzekou mission was being burnt to the ground. A villager told them that the lamas had set an ambush for them on the trail ahead, so that they were blocked off from reaching Yetche. More Tibetan tribesmen were racing at them from the north.

'East down a breakneck path'

George Forrest never wrote his promised memoir, but he did pen a long letter to a friend and from this we have a vivid account of what happened next, as the fugitives found themselves trapped on a trail on the banks of a stream that was a tributary of the Mekong River:

> *To the north I had a clear view of the crest of the ridge we had descended ... Suddenly there appeared a large number of armed men running at full speed in Indian file along the path we had just traversed. I gave the alarm at once and immediately all was confusion, our*

followers scattering in every direction. Pere Bourdonenec [one of the elderly priests] became completely panic-stricken, made his way across the stream by a fallen tree, and, despite my efforts to stop him, rushed blindly through the dense forest which clothed the southern face of the valley ... the Pere had not covered a couple of hundred yards [metres] ere he was riddled with poisoned arrows and fell, the Tibetans immediately rushing in and finishing him off with their huge double-handed swords. Our little band, numbering about 80, were picked off one by one or captured, only 14 escaping. Of my own 17 collectors and servants only one escaped.

Forrest quickly assessed the situation. They were in a valley that was really a small rift in the mountains. It was about two and a half kilometres (one and a half miles) wide, hemmed in by the Mekong River to the east and, to the west, north and south, high ridges that were now covered with armed and enraged Tibetans. There was only one way for Forrest to go:

I fled east down a breakneck path, in places formed along the faces of beetling [overhanging] cliffs by rude brackets of wood and slippery logs. On I went down toward the main river, only to find myself, at one of the sharpest turns, confronted by a band of hostile and well-armed Tibetans, who had been stationed there to block the passage. They at once gave chase. For a fraction of a time I hesitated; being armed with a Winchester repeating rifle, twelve shots, a heavy revolver and two belts of cartridges, I could easily have made a stand, but I feared being unable to clear a passage before those I knew to be behind me arrived on the scene. Therefore I turned back and, after a desperate run, succeeded in covering my tracks by leaping off the path whenever I rounded a corner. I fell into dense jungle, through which I rolled down a steep slope for a distance of 200 feet [60 metres] before stopping, tearing my clothes to ribbons and bruising myself horribly in the process.

Frank Kingdon-Ward

In the summer of 1922 George Forrest was 'irked', as he wrote to a friend, to discover that there were two other plant collectors in the area of Yunnan that he considered to be 'his' territory. One of them was an American named J. F. Rock. The other was Frank Kingdon-Ward, whom Forrest considered to be his main rival, since he had been hired by Forrest's old employer, A. K. Bulley.

Frank Kingdon-Ward was one of the finest plant collectors in China in the early part of the twentieth century. Born in 1885, he was the son of a professor of botany at Cambridge, Harry Marshall Ward, and was himself educated there, before moving to Shanghai to take up a post as a teacher. Numerous plant-collecting side trips into the Chinese countryside, during which he would send back plant specimens to Cambridge, convinced him that botany was his main vocation. Much like Forrest before him, Kingdon-Ward got his break when A. K. Bulley asked Sir Isaac Bayley Balfour for a plant collector, since Forrest had now signed on with the Williams syndicate. Kingdon-Ward reached an agreement with Bulley in 1911 and began his career as a plant collector, one that lasted all the rest of his seventy-three years—he was, in fact, planning a new expedition when he died in 1958.

Only three of Kingdon-Ward's numerous expeditions took place in China, including the 1922 foray that so irritated Forrest. Most of Kingdon-Ward's plant hunting took place in Upper Assam, Burma and Tibet, where he managed to get out of several close scrapes. He sent back numerous new species to Bulley, including Rhododendron wardii *and* Primula burmanica. *In 1924–25 he took an extraordinary journey deep into the Himalayas in Tibet, finding the rare* Rhododendron leucaspis—*its 'budscales finely-fringed with hairs like spun glass'.*

During World War II Kingdon-Ward had more adventures, working as a Special Operations officer to guide troops through the wilds of Burma, and after the war helping to search for Allied aircraft that had been lost in the jungle.

'One more bid for life'

Forrest hid behind a boulder and prepared to make a last stand, certain the Tibetans would discover his hiding place, but they rushed by, up the valley, thinking that he had continued on in that direction. When nightfall came, he climbed back up towards the southern ridges, hoping to make his escape, but found his way blocked 'by a cordon of Lamas with watch-fires and Tibetan mastiffs'. He returned to his hiding place. For the next eight days he slept during daylight hours and crept out at night, hoping to make it south, but he could not get past the Tibetans and their dogs. All he had to eat was a handful of parched peas, which he assumed had been dropped by a fugitive or perhaps by a Tibetan tribesman. He had taken off his boots in order to avoid leaving Western tracks. Parties of Tibetans sought him everywhere. One time he was shot at without warning, with two poisoned arrows passing through his hat. He managed to escape. Another time, as he slept in a log in a streambed, he was awoken by thirty lamas passing by in single file 'in full war paint'.

'At the end of eight days,' Forrest wrote, 'I had ceased to care whether I lived or died; my feet swollen out of all shape, my hands and face torn with thorns, and my whole person caked with mire. I knew the end was near and determined to make one more bid for life.' Knowing that there was a village nearby, Forrest decided on a desperate gamble: he would hold up the town with his superior firepower and take their food. But it turned out the village was inhabited by a 'sub-tribe' of Tibetans, the Lissu, who did not sympathise with the lamas' rebellion. Instead, at great risk to himself, the headman gave Forrest food and shelter, and guided him to the Mekong, where he was handed over to another headman. Tibetan search parties were still looking for him—he just missed one as he approached the second headman's village—and he knew he could not stay in the region much longer.

There was no question of going directly south, since that way was completely blocked off by Tibetan forces, and so it was decided that

Forrest had to make a complete circle around the dangerous area, which meant climbing the mountains to the west. The journey was astonishingly hard. It was the middle of the rainy season and 'I soon found myself in the thick of the worst downpour Yunnan had known for a generation.' Then they began climbing the mountains. 'Up and up we climbed,' Forrest wrote, 'cutting our way through miles of Rhododendrons, tramping over alps literally clothed with Primulas, Gentians, Saxifrages, lilies, etc., til we reached the snowfields on the backbone of the range, at an elevation of from 17,000 to 18,000 feet [5000–5500 metres].'

It was torture for the plant hunter to pass these untouched treasures and not take specimens, but he pressed on, his feet torn by rocks and ice. They turned south atop the ridges and marched for six days 'over glaciers, snow and ice, and tip-tilted, jagged limestone strata'. Finally, his guides felt it was safe for them to descend east towards the Mekong and the town of Yetche. Forrest would have one more unpleasant surprise—he stepped on a sharpened *punji* stick a peasant had put out to protect his crops and pierced his foot through—but he finally escaped. He had only his rifle, pistol and ammunition, and 'the rags I stood in'.

Forrest would mournfully count his losses—'2,000 species [of plants], 80 species of seeds, 100 photographic negatives'—but he had his life. His old host Father Dubernard was not so lucky. Captured after hiding for two days in the woods, he had been strapped to a post, his eyes were gouged out and his nose, ears and tongue were cut off. He lived for three days. On the third day he was disembowelled alive, and then beheaded and quartered.

'The deep blue sky above'

Forrest finally returned to Tengchong at the end of September, where his friend, the consul George Litton, helped nurse him back to health. However, such was Forrest's incredible hardiness, as well as his resilience of spirit, that by 11 October he was ready for another adventure—this time a non-botanising expedition with Litton up the Salween Valley, to an area where no European had ever been. The locals there were a tribe known as the Black Lissu, who were reputed to be as opposed to foreign encroachment as the Tibetans. (A few months after Litton and Forrest passed through, the tribe murdered a party of Germans who had entered their territory.) The going was dangerous. At one point they were approached by a tribal prophet of the Black Lissu, who claimed that he had received advice in a dream that he should kill someone and was on his way to attack a neighbouring tribe and murder its chief. But he wanted to know what Litton and Forrest thought of this. Forrest advised him to go home and tend to his crops. To their astonishment, the man took their advice.

Still, the wilderness beauty around them was incredible and Forrest's later article in *Geographic Journal* was worthy of it. He describes climbing one mountain:

> *After an intensely cold night on the mountain-side at 10,500 feet [3200 metres] we proceeded on November 19th up the pass, which for the first time was traversed by European feet. The path, after topping a spur, led through the pine woods deep in snow, and then over a frozen black marsh surrounded by tall sombre firs, whose dark green foliage stood out against the snowy slope of the pass and the deep blue sky above.*

Forrest and Litton returned to Tengchong in mid-December. Unfortunately, George Litton had contracted blackwater fever, a type of malaria, and died a month later. Forrest himself had been exposed to Salween malaria, a particularly virulent version of the disease that did not, however, begin its onset until the following summer. In March 1906 he decided to botanise in the Lijiang Mountains, the vast range that forces the Yangtze in a great loop to the north. In the coming years, these mountains would become Forrest's main flower-gathering area; he described the place as 'a huge natural flower garden', some 80 kilometres (50 miles) in length, from the mountain valleys up to where the vegetation ended at about 5200 metres (17,000 feet).

In August, however, the Salween malaria hit Forrest and he became deathly ill. He was forced to return to the town of Talifu, where the British doctor told him he needed to return to Britain if he wanted to survive. This Forrest refused to do. He had already lost one collecting season after his perilous experience with the Tibetans; he did not want to lose another one. Fortunately, he had trained his collectors so well that, as he said, 'I had much reason to be grateful to them'. A system of runners had been set up, so that plants were brought to him as quickly as possible—the mountains were a nine-day walk away from Talifu—and he lost only a few specimens. He was able to ship back to Bulley in Britain over a hundred kilograms of seed and numerous plant specimens, including one fine primula he named *Primula bulleyana*—after his sponsor, of course. It began flowering in Britain some three years later and has thrived there ever since.

Rhododendron sinogrande

Forrest's most successful expedition was his third journey, between 1912 and 1915. He had returned to Great Britain in 1911 and begun working for a large syndicate backed by J. C. Williams of Caerhays Castle in Cornwall, England (although Bulley may have been, for a time, a minority partner in this group). Arriving in China in March 1912, Forrest stepped into a revolution, as the Qing Dynasty had been toppled and various factions fought for control of the land. Bandits and corrupt warlords ruled large portions of the countryside. Forrest found that Tengchong had been taken over by revolutionary forces and that a perfect reign of terror was occurring—'beheading is the order of the day', Forrest wrote in the first of a series of letters to J. C. Williams. There were numerous other difficulties—a new form of currency had been issued, making it difficult for Forrest to get merchants to accept his cheques, and the price of food had skyrocketed.

Nonetheless, Forrest set off with his trained collectors into the wilds around the Lijiang Mountains. Forrest did extremely well with the Chinese. He spoke the language, as well as several local dialects; he could doctor and dispense medicine (he once had an entire group of his collectors and their families inoculated for smallpox); and he treated them patiently and fairly, which in general was not their experience with Europeans. They responded with extreme loyalty. By July Forrest was able to write that he had thirty-five seed-drying presses going and that 'the collection is steadily increasing at the rate of 150 specimens per week. I now have 1,100 specimens in hand, 850 of which are dried, numbered, written up and packed.'

However, in August the revolution in Yunnan grew hotter, with bandits, revolutionaries and soldiers of the local warlord roaming the roads, and much as he hated it, Forrest thought it was safer for him to 'retire' to Burma, leaving his plant collectors, who were not in as much danger as a European, behind to continue to work. 'I am extremely sorry to have to inform you that serious trouble has arisen near here and I may have to get out at any moment ... I think I told you before I was a Jonah.'

By 1913 things had quieted down and Forrest was able to return to Yunnan. Heading back up into the mountains, he came upon one of the most significant finds of his career, the plant that would become known as *Rhododendron sinogrande*. In 1905, while escaping from the Tibetans over the mountains, he had noticed to his astonishment rhododendrons growing out of limestone rocks—something that was not thought possible because the lime should have killed the roots of the plants. Forrest was the first to realise that the lime contained in the limestone rock is unlike the type of lime found in alkaline soil, and thus it does not harm plants.

The rhododendron he found in November was 'a magnificent species', as he wrote to Williams. 'The capsules are 2–2½ inches long, slightly curved, and as thick as one's thumb. The foliage runs from one foot by six inches to as much as two feet by ten inches

[45–85 centimetres], dark green and glossy on the upper surface, ash coloured beneath. Very handsome tree of 20–30 ft [6–9 metres]. More later.'

Some subspecies of *Rhododendron sinogrande* grow to as high as 12 metres (40 feet) in the wild. Forrest, on this and subsequent trips, sent back numerous seeds of *Rhododendron sinogrande* and one was finally grown successfully in an English greenhouse in 1919.

Upon his return from the mountains to Talifu in December of 1913, George Forrest's Jonah-like bad luck held. On the day he entered the city, the soldiers garrisoned there rebelled and killed their commanders. Forrest was held prisoner there for three weeks and was forced to act as a doctor to wounded rebels. The city was finally retaken by loyal troops in a bloody battle on 24 December. 'Order being restored once more,' Forrest writes, with an equanimity born of much exposure to such upheavals, 'I proceeded on my journey to Tengyueh [Tengchong], packed up my collections there, and despatched the whole to England.'

After that, he went back up into the mountains.

'Primulas in profusion'

For the next two decades, George Forrest roamed the mountains of Yunnan; in all, his plant expeditions covered twenty-eight years. He brought back to Great Britain three hundred new species of rhododendrons, and thirty thousand plants in all. Although he now had a wife and three sons, he could not keep himself from going back to China for expedition after expedition. Each time he went back he said that the expedition he was beginning would be his 'last'. In 1930 he decided on one more 'last' expedition, during which he would visit all his old spots in Yunnan as well as try to fill in gaps in his previous collections. It was a satisfying trip, he reported, early in 1932:

> *Of seed such an abundance, that I scarce know where to commence, nearly everything I wished for and that means a lot. Primulas in*

profusion, seed for some of them as much as 3–5 lbs [1.3–2.2 kg] ... When all are dealt and packed I expect to have nearly if not more than two mule-loads of good clean seed, representing some 400–500 species, and a mule load means 130–150 lb [60–70 kg]. That is something like 300 lb. [136 kg] of seed. If all goes well, I shall have made a rather glorious and satisfactory finish to all my years of labour'.

On 5 January 1932, shortly after he wrote this letter, George Forrest was out hunting for birds when he suddenly gasped, grabbed his chest and fell dead to the ground from a heart attack. He never did return from China, being buried next to his friend George Litton in the graveyard at Tengchong.

ROY CHAPMAN ANDREWS
Dinosaurs in his bones

The Gobi Desert of southern Mongolia stretches 1600 kilometres (1000 miles) from east to west and 1000 kilometres (600 miles) north to south, creating a breathtaking moonscape of salt flats, endless marshes, sand dunes, and grotesque cliffs and rock formations. Its main feature, however, is more desolate than dramatic—hundreds of kilometres of windswept plains covered with *gobi*, the Mongolian word for gravel.

The Gobi is the second largest desert in the world—only the Sahara is larger—but no desert on earth has such temperature swings. In the summer, temperatures range as high as 60° Celsius (140° Fahrenheit) in the sun and 45° (110°F) in the shade. In the winter—as early as October, as a matter of fact—the Gobi can get to minus 45° (minus 50°F) and howling blizzards are not uncommon. These latter are not to be outdone by sand storms, one of which was described by a traveller named Roy Chapman Andrews in 1922:

> *It came like a cyclone bringing a swirling red cloud of dust. In less than ten minutes the temperature dropped at least thirty degrees. A thousand shrieking demons seemed to be pelting my face with sand and gravel ... We could not see twenty feet [6 metres], but we heard the clatter*

*of tins, the sharp rip of canvas, and then a tumbling mass of camp
beds, tables, chairs, bags, and pails swept down the hill. Lying flat on the
ground with our faces buried in wet clothes we at least could breathe.*

As the twentieth century began and revolution convulsed China
and Russia, the Gobi was home mainly to wandering Mongolian
tribespeople, descended from Genghis Khan and his original
Mongols, who had once been masters of the desert and the steppes
beyond it. Aside from them, the nearly empty land held a few hardy
desert sheep, circling buzzards, a strange and reclusive species of
desert bear, and thousands of extremely poisonous brown pit vipers.
That is, until Roy Chapman Andrews literally burst upon the scene as
the Roaring Twenties were about to begin. Leading an incongruous
procession of Dodge motorcars (whose windshields the howling
Gobi wind immediately shattered) Chapman was on an extraordinary
hunt for something even more breathtaking then the Gobi itself: the
fossilised bones of the oldest creatures on earth.

'Like a rabbit'

It has long been rumoured that Roy Chapman Andrews was the
model for the dashing movie archaeologist–adventurer Indiana Jones,
and while this does not appear to be directly true, one can see why
people might think so. Roy Chapman Andrews was a palaeontologist
and zoologist rather than an archaeologist and carried a 6.5-mm
Mannlicher carbine instead of a rawhide whip. During the course
of five major expeditions to Central Asia between 1922 and 1930 he
battled bandits, civil war, blizzards and sandstorms—as well as the
scoffers of the scientific establishment—to make palaeontological
history. The bones Andrews and his fellow expeditionary members
discovered in the Gobi became the species of dinosaur we are familiar
with today—*Velociraptor* and *Protoceratops*, just to name two—while
the fossilised eggs he found at the ethereally beautiful Flaming Cliffs
confirmed to science that dinosaurs hatched their young.

Real life meets fiction?

Dr Henry Walton 'Indiana' Jones, Jr, is a movie character—created by American director George Lucas and played by actor Harrison Ford—who first appeared in the 1981 movie Raiders of the Lost Ark *and has since been featured in four other major feature films, as well as television shows. Jones even has his own amusement park attraction at Disneyland. An adventuring archaeologist who finds himself in numerous scrapes in exotic places, he is long, lean and handsome, carries a bullwhip, wears a fedora and a leather jacket, and has a serious phobia about snakes.*

The American public of the 1980s worshipped Indiana Jones immediately, but many of those fans were too young to remember Roy Chapman Andrews, to whom Jones bears more than one passing resemblance—there's the physical one, a signature weapon and that fear of snakes. Did Lucas base Jones on Roy Chapman Andrews? Many people assume he did, although Lucas himself has never admitted this. Interestingly, America's Smithsonian television channel, which specialises in popular science and history, has looked into this question and discovered that Hollywood directors such as George Lucas and Stephen Spielberg based their adventure characters (and movies like Raiders *and* Star Wars) *on characters in old 1940s movie serials—which were in turn inspired by articles and books written by Roy Chapman Andrews. So the linkage between Jones and Andrews is real, if indirect.*

Roy Chapman Andrews was born in Beloit, Wisconsin, in 1884, the son of a middle-class druggist and a bookish mother. Growing up in small town America near the turn of the century was a pleasant, if somewhat narrow, experience, and it suited Andrews well. He loved to roam the woods and fields of Wisconsin, where he quickly became an adept naturalist. 'I was like a rabbit,' he later wrote, 'happy only when I could run out of doors'. By the time he was in his teens, Andrews had taught himself to become a taxidermist. He was so skilled that he was licensed by the state to start a part-time business; according to Charles Gallenkamp, Andrews's chief biographer,

Andrews's talents 'were solicited by family friends and schoolmates who brought him their dead pets to be mounted—everything from dogs, cats, and parrots to snakes, lizards and turtles'.

At the age of seventeen, this precocious young man—tall, lean, with a jaw that looked carved with a sculptor's chisel out of solid rock—went to Beloit College to seek a Bachelor of Arts degree. The drowning death of a good friend—out on a canoeing expedition with Andrews, who barely escaped the accident with his life—made him realise the transitory nature of existence and he vowed to let nothing stand in the way of his dreams. For years he had wanted to work for the American Museum of Natural History—'Nothing else ever had a place in my mind,' he later wrote—and upon graduation in 1906 he headed to New York with a train ticket bought for him by his father, and thirty dollars, his life savings, in his pocket.

A foot in the door

The Museum of Natural History, on Central Park West at 79th Street in New York City, was founded in 1869 by a host of American dignitaries, including Theodore Roosevelt, Sr, father of the twenty-sixth president of the United States; financier J. Pierpont Morgan; eminent American historian Charles Dana; and naturalist Dr Albert S. Bickmore, whose dream the museum had been. The purpose of the museum was to house one of the finest natural history collections in the world—this was a period when the taking of bird, reptile, mammal, fish, insect and botanical specimens was considered the *raison d'être* of the natural sciences. But very quickly, under its charismatic president Morris K. Jesup, the museum launched a virtual golden age of exploration, sending groundbreaking expeditions to places ranging from the North Pacific to the densest jungles of the Congo. These were written up in the popular press and became the subject of books by expedition members, and soon the museum had—in the eyes of the American public—become the world's most prestigious monument to the pursuit of the natural sciences.

Thus, when Roy Chapman Andrews alighted from his train in the summer of 1906, it was only natural that he would head straight for its august doors. What followed has become a famous part of the Andrews legend. The lanky twenty-two-year-old secured an appointment with Hermon Bumpus, director of the museum. Despite the fact that Bumpus was firm in telling the young man that no jobs were available, Andrews persisted, saying that he would mop floors, do anything, just to be able to work in the celebrated museum. Impressed by Andrews's ardour, Bumpus hired him as an assistant in the taxidermy department at the grand salary of forty dollars per month.

It was a foot in the door, and it was all Roy Chapman Andrews needed. His knowledge of taxidermy and animal anatomy got him the job of making a life-sized papier-mâché model of a whale for an exhibition. After that, he began gathering whale skeletons, travelling to New England, British Columbia, Alaska, Japan and Korea. Between trips, he began to study mammalogy at Columbia University, although staying in one place for too long made him restless.

'I was born to be an explorer'
In 1909, on a trip to the Far East, Andrews—a collector of exotic human beings as well as animals—befriended a famous Japanese geisha madam known as Mother Jesus ('attractive, with calm appraising eyes behind which seemed to lie the wisdom of the ages') and then sailed to the Philippines, where he survived a typhoon as well as being stranded on a desert island for nearly a week. Rescued and brought to Manila, Andrews joined a scientific expedition sponsored by the US Navy and travelled throughout the islands of the Philippines, the South Pacific and Southeast Asia. In Buru, he dodged hostile tribesmen who had set traps for the expedition using poisoned stakes. In Borneo, he killed a 6-metre (20-foot) python slithering towards him. The only disappointment of the entire expedition was being unable to go ashore in Formosa to collect

animal specimens—head-hunters were on the prowl, Andrews was told, making it far too dangerous.

All in all, however, the trip was, as Andrews would write, a journey to 'places of pure enchantment', one that whetted his appetite for more. While he was supposed to return home in January of 1910, he secured the museum's permission to stay in Japan to study whales (the Japanese whaling industry at the time was one of the most important in the world) and from there found his way to China, visiting the country just as the old Manchu regime was crumbling and the revolutionaries of Sun Yat-sen were coming to the fore. Andrews was not politically inclined, however, and, as Charles Gallenkamp reports, paid more attention to the caravans coming from western China, long lines of snorting camels 'laden with goods from Tibet, Mongolia, Russia, Afghanistan and India'.

Finally making his way home via the Middle East, Andrews arrived back in New York in January of 1911. His former mentor, Hermon Bumpus, had resigned, but Andrews soon became the protégé of the president of the museum, Henry Fairfield Osborn, who had replaced Morris Jesup after the latter's death in 1908. Osborn, a scientist and outsized personality, listened to Andrews's enthusiastic tales of Asia and backed him on another expedition, this time to Korea, where Andrews wanted to seek the so-called 'devil fish', a whale with strange flukes that Andrews suspected was actually the California grey whale, which scientists thought had been hunted into extinction.

Following whalers in Korea the next year, Andrews actually proved his point—far from being extinct, the grey whale was alive and well in Asia—and he came home with pictures and a whale skeleton to prove it. To cap off his successful trip, he took a side-trip to the Paik-tu-san (Long White Mountain) a 2700-metre (9000-foot) peak in northern Korea surrounded by wilderness that only a few people had ever penetrated.

The outsized Henry Fairfield Osborn

In some ways an even more extravagant character than Roy Chapman Andrews, Henry Fairfield Osborn was born in Fairfield, Connecticut, in 1857, to a wealthy family—his father was shipping magnate William Henry Osborn and the tycoon J. Pierpont Morgan was his uncle. Osborn became a professor of biology and zoology at Columbia University, studied palaeontology and eventually rose to take over the presidency of the American Museum of Natural History.

Osborn was influential in sending Andrews and other natural explorers out into the world on much publicised expeditions and thus raising the stock of the museum in the eyes of the American public. He was, however, prone to a certain vanity. An imposing man with a large head of thick grey hair, he arrived at work each morning in a chauffeur-driven limousine and headed to the fifth floor of the museum, where he had established the Osborn Library, in which the chief feature was a bust of himself highlighted by a single spotlight. He belonged to 158 scientific organisations and his masterpiece of scholarship was his two-volume book Proboscidea: A Monograph on the Discovery, Migration and Extinction of the Mastodonts and Elephants of the World. *It took him over half a century to write, weighed 18 kilograms (40 pounds) and cost 280,000 dollars to produce (money provided by J. P. Morgan) when it was published in 1936, a year after Osborn's death.*

Osborn was a distinguished personage of his day, but his legacy has been marred by racist beliefs that made him unable to accept that humankind could have originated in Africa.

When Andrews returned to New York, he realised that he had become a celebrity as reports spread of his travels through Asia and how he had reclaimed the grey whale from its watery tomb of extinction. He began to write and lecture a good deal, gaining a measure of fame that never really left him. He gained, as well, the jealousy of certain members of the scientific establishment, who whispered that he was not a serious naturalist and was, in fact, out to make money and a name for himself rather than further the ends of science. It was a charge that would follow Andrews all his life and there was some justice to it, although it was untrue that Andrews ever did what he did solely for fame or money. Despite the Master of Science degree he would earn from Columbia in 1913, he was impatient with the painstakingly detailed work of collecting specimens (he had what one historian has described as 'a casual approach to science'). Andrews was a populariser, a restless traveller and adventurer, rather than a pure scientist. In his memoirs he wrote: 'I was born to be an explorer ... I couldn't do anything else and be happy.'

The Asiatic Zoological Expedition

After his last trip to Korea, Andrews began to focus more and more on land expeditions, rather than those involving whales and other sea creatures, in good part because he became fascinated with an argument that Henry Osborn and a few other scientists were putting forward at the time, to wit, that northeast Asia was the area where mammals and humans first originated—the 'Garden of Eden', as it were, of warm-blooded life on this planet. 'The main problem,' Andrews wrote, 'was to discover the geologic and palaeontologic history of Central Asia, to find out whether it had been the nursery of the dominant groups of animals, including the human animal.' To do this, he further explained, 'it was necessary that a group of highly trained specialists be taken into Central Asia'.

When Andrews pitched the idea of a grand series of expeditions to Asia to Henry Osborn, the museum president gave him his blessing, if not his entire financial backing. For the first of the journeys of what would be called the Asiatic Zoological Expedition, Osborn would only advance Andrews half of the fifteen thousand dollars needed. The rest Andrews would have to raise himself from prestigious donors such as industrialists Henry Clay Frick and Sidney M. Colgate. This was not hard for the dashing, charming

Andrews, especially since he had just married the lovely Yvette Borup, sister of a good friend who had recently died tragically in an accident. Yvette, as well as having connections in society, was quite adventurous and joined Andrews in the spring of 1916 when he embarked for Yunnan Province in southwestern China for an initial exploratory expedition along the wild Tibetan frontier, a place no zoologist had ever explored, although George Forrest, the plant hunter, was finding this same area fertile ground (see page 286). Reporters interviewing the couple as they boarded an ocean liner in San Francisco shouted questions about Yvette—would she wear a dress? carry a rifle? To which Andrews replied that she 'would wear exactly what the men wore ... and would indeed carry a rifle and not hesitate to use it'.

The American public ate it up. Andrews and Yvette became fodder for the tabloid press and soon for the silent newsreels that the burgeoning American film industry had begun to display. The couple were romantic heroes, yet their adventures were real enough. Once in China, they were caught up in the civil war between rival revolutionary factions and, wearing crude red crosses pinned to their chests, acted as stretcher-bearers for the wounded. After this, they travelled overland to the Tibetan border, taking literally thousands of specimens of local wildlife, enough to fill forty-one huge crates shipped back to the Museum of Natural History. Back in America over a year later, with Yvette pregnant with their son, Andrews furthered his exotic legend by enlisting in the spring of 1918 as a spy for the US government, which had declared war against Germany. Given a code name, Andrews was sent to China on another expedition, but this time with the secret mission of spying on the Russians, uneasy US allies now that the Bolsheviks had taken over.

It was on this latest journey to China, in August of 1918, that Andrews slipped over the border for the first time into Mongolia. Riding horses, led by Mongolian guides, he found himself in an extraordinary land. He later wrote:

Never again will I have such a feeling as Mongolia gave me. The broad sweeps of dun colored gravel merging into a vague horizon, the ancient trails once traveled by Genghis Khan's wild raiders ... All this thrilled me to the core. I had found my country. The one I was born to know and love.

After the war was over, Andrews returned to Mongolia with Yvette, visiting the capital city of Urga (now Ulaanbaatar), whose ancient beauty—it was founded in 1649—enthralled them. At the same time, they were introduced to the savagery of life in Mongolia, as a local official showed them the barbaric jails where manacled prisoners were forced to crouch in boxes too small for them to stand up in, sometimes for years at a time.

It was during this expedition Andrews found what he suspected to be fossilised rocks and knew instantly that this should be the centre of his first major Asiatic Zoological Expedition.

'The fossils are there'

In New York in the spring of 1920, Andrews had a lunch with Henry Osborn and told him that he intended to focus his hunt for fossils (and early humans) in the area of the Gobi, which was practically unexplored by scientists. Not only that, but he had a special plan for doing so. His expedition of scientists would be carried by Dodge cars and Fulton trucks—the fruit of the assembly line revolution in the American auto industry—which would be supported by at least dozens of camels carrying gasoline and spare parts for the cars and supplies for the expedition. Since one could not work in the Gobi in the winter, expeditions would take place in the summer. The entire enterprise would be based in Peking (Beijing).

Andrews had his plan highly organised. Each expedition sent forth would be divided into three or four autonomous units, each with scientists (palaeontologists, geologists, zoologists and anthropologists), a small fleet of automobiles and supplies and fuel for two weeks so that if they got separated—a likely event given the rough terrain and dramatic weather of the Gobi—they would be self-sustaining.

Andrews told Osborn that the whole venture might take as many as five years but, when they were done, they would have an extraordinarily complete fossil picture of the Gobi. 'We should try,' Andrews declared, 'to reconstruct the whole past history of the Central Asian Plateau.' Osborn agreed enthusiastically—the magnificent adventure appealed deeply to him—and together, he and Andrews set about raising the funds necessary to finance the expedition. Two hundred and fifty thousand dollars was found almost immediately, fifty thousand dollars donated by banker J. Pierpont Morgan. The expedition caught the interest of the American public, and thousands of telegrams were sent to Andrews by young men and women who were eager to take join the expedition. However, publicity had its downside. When Andrews said he hoped to find the fossils of near-humans—the so-called missing link between animals and the first man and woman—the tabloid press seized on it with a vengeance. Headlines blared that Andrews was on the hunt for 'ape-men' or 'dawn-men', and Andrews received racist letters, including one that read: 'Why go to Asia for the missing link? I saw one in the subway this morning.'

There were critics within the scientific community, as well. There had only been one fossil found in the Gobi Desert—the tooth of a prehistoric rhinoceros discovered in 1894—and critics claimed that Andrews 'might as well look for fossils in the Pacific Ocean as in the Gobi'. None of this deterred Andrews. In the winter of 1921, Henry Osborn sent what was officially known as the Third Asiatic Expedition off with an almost biblical injunction: 'The fossils are there. I know they are. Go and find them.'

'You have written a new chapter'

On 21 April 1922 the expedition passed through a gate in the Great Wall of China at the town of Kagan, 160 kilometres (100 miles) northwest of Peking. Entering Mongolia, Andrews wrote, 'the hills swept away in the far-flung graceful lines of a vista so endless that we seemed to have reached the very summit of the earth'. Andrews's lead group consisted of seventy-five camels, three Dodge touring cars and two Fulton trucks, which made 430 kilometres (265 miles) in four days, an unheard of distance in the Gobi. Mongolian herdsmen were astounded by the cars, which most of them had never seen before, and gaped as they roared past.

Andrews had with him Yvette, whom he was going to drop off in Urga (by decree of the Board of Directors of the museum, no wives were allowed along on the expedition), as well as several distinguished scientists, including a Canadian anthropologist named Davidson Black and, notably, the American Walter Granger, who served as chief palaeontologist for the expedition. Within a week of beginning their travels, Andrews was told by Mongols that there was an area where 'bones were to be found, bones as big as a man's body'. He sent Granger off to investigate while the main party travelled on to make camp for the night. In his later account of the expedition, Andrews wrote:

> We were hardly settled before Granger's car roared into camp. The men were obviously excited when I went out to meet them. No one said a word. Granger's eyes were shining and he was puffing violently at his pipe. Silently, he dug into his pockets and produced a handful of bone fragments; out of his shirt came a rhinoceros tooth, and the various folds of his garments yielded other fossils. He held out his hand. 'Well, Roy, we've done it. The stuff is here. We've picked up fifty pounds of bone in an hour.'

It was only the beginning. In August, in a 'vast, pink basin', they found 'caverns [running] deep into the rock, and a labyrinth of ravines and gorges studded with fossil bones [that] make a paradise for the paleontologist'. One scientist present simply walked up to a mound of sandstone and found a fossil identified as an ancient reptile. Other fossilised bones, most from the Cretaceous period, could simply be picked up at will. Looking across the pink-walled canyon as the sun was setting, Andrews found himself awed by the explosions of red and orange and called the area the Flaming Cliffs, a name that would make the place famous throughout the world.

The ages of the dinosaur

The Mongolian herders who met Roy Chapman Andrews referred to him as 'dragon-hunter' because they believed that the bones lying scattered around the Gobi Desert belonged to these mythical creatures. In fact, they belonged to animals even more ancient, but certainly not mythic: the dinosaurs.

When dinosaurs first appeared, in what is known as the Triassic Era (from 248 to 213 million years ago), the earth is thought to have been a single landmass, called by geologists Pangaea, the united earth. As the land gradually separated and continents began to form, dinosaurs lived through the Jurassic (213 to 144 million years ago) and Cretaceous (144 million to 65 million years ago) periods, after which they disappeared after millions of years on earth, leaving the way clear for mammals—including, eventually, humans—to flourish. (The reason for the disappearance of the dinosaurs has long been debated; many scientists now believe that a large meteorite hitting the earth, or several massive volcanic explosions, caused dust to fill the earth's upper atmosphere and begin a severe cooling period.)

Whatever happened, millions of dinosaurs left behind numerous bones that became fossilised and thus preserved, and humans have been wondering at them ever since. During one of his digs in Mongolia, Roy Chapman Andrews discovered a Stone Age site where humans had worked dinosaur eggs into patterns—'These people were the original discoverers of the dinosaur eggs,' he wrote. North American Indians described dinosaur bones as 'the grandfather of the buffalo' and Europeans of medieval times found biblical explanations for the bones. It was only in the mid-nineteenth century that early palaeontologists began a systematic study of these fabulous creatures, making the science that Roy Chapman Andrews pursued a young, cutting-edge one.

The scientists set to work with a will, excavating the remains of numerous dinosaurs and even, as Andrews wrote, 'parts of the skeleton of *Baluchitherium*, the largest known land mammal'. Andrews directed the digs but was not terribly patient about the painstaking process of extracting the fossils from solid rock (other scientists referred to his cracked fossils as having been 'RCA'd').

When September came, the freezing winds of the Gobi began howling across the gravel-strewn plains. Reluctantly, the scientists gave up their search and returned to Peking, arriving there five months after they had set out. They had been out of touch with the outside world for the entire period and Andrews now sent a cable to the museum and Henry Osborn, enumerating his triumphs, which included, as he wrote, 'the discovery of complete skeletons of small dinosaurs, and parts of fifty good dinosaurs, and ... [even] wonderfully preserved Cretaceous mosquitoes, butterflies, and fish'.

Osborn replied by telegram, with typical hyperbole: 'You have written a new chapter in the history of life upon the earth.'

The eggs

Despite the fact that they had not discovered the missing link, Andrews knew that the expedition had found something of major importance and returned to Mongolia, and the Flaming Cliffs, with some impatience in April 1923. The country, still claimed by both Russia and China, was essentially lawless. Andrews's expedition travelled at first with the protection of Chinese soldiers, but soon these returned behind the confines of the Great Wall, leaving the heavily armed members of Andrews's expedition (the Dodge touring cars, he wrote, 'were bristling with rifles') to defend themselves. At one point, returning alone from a resupply mission, Andrews encountered three bandits. After a few shots with his Colt .38 failed to stop them as they charged at his car on their ponies, Andrews took more drastic action:

The cut-out was wide open and, with a smooth, down-hill stretch in front of me, the car roared down the slope at 40 miles [65 kilometres] an hour ... While the bandits were attempting to un-ship their rifles, which were slung on their backs, their horses went into a series of leaps and bounds, madly bucking and rearing with fright, so that the men could hardly stay in their saddles.

Andrews's unexpected charge frightened the bandits' ponies so much that the last he saw of them they were riding off in a frenzy over the hills. Shortly after this encounter, Andrews and his increased crew of palaeontologists—who included not only of Walter Granger, but also two Danish experts, Peter Kaisen and George Olsen, as well as an American fossil-hunter named Albert Johnson—stumbled on an extraordinarily important find. On 13 July, deep in the basin of the Flaming Cliffs, Olsen told Andrews that he had discovered a clutch of dinosaur eggs. At first, none of the others believed him. Although palaeontologists at the time surmised that dinosaurs did indeed lay eggs, there had been no proof of this. But then Olsen led Andrews and the others to a small sandstone ledge. And there, indeed, 'were lying three eggs partly broken. The brown shell was so egg-like that there could no mistake'. The eggs measured about 200 by 180 millimetres (8 by 7 inches); the shells were corrugated.

It was the first time intact dinosaur eggs had been discovered. (Egg fragments had been found in France in the mid-nineteenth century but were not recognised as such until the mid-twentieth century.) To make matters more astonishing, more clutches of eggs were found—five here, nine there, thirteen in two layers in a block of sandstone. Some eggs were cracked and contained what Andrews and his colleagues thought were dinosaur embryos. (In the 1960s these were identified as calcite crystals, although dinosaur embryos have since been found.)

These eggs, which had never hatched and may have been buried in sandstorms, were probably the eggs of the dinosaur *Protoceratops*, numerous skeletons of which had been found by Andrews and his men. But other exciting fossils were discovered as well. One was the skeleton they named *Saurornithoides*—the 'bird-like reptile'. Another was *Velociraptor*—the 'fast-running robber' that Andrews envisioned as plundering the eggs of *Protoceratops* and that was a truly frightening feathered predator with a sickle-shaped claw on each hindfoot, the better to eviscerate its prey. (The *Velociraptor*, or raptor, later became infamous for its starring turn in the movie *Jurassic Park*.)

Some of the finds made life during the Cretaceous period come leaping to life. At one point, Granger and another expedition member saw, sticking straight up out of the earth, the whitened leg-bone of a *Baluchitherium*. They could not figure out how it came to be left this way, until they found the other legs at intervals of about 2.7 metres (9 feet) and realised that the animal had been caught in quicksand.

There were dangers for modern humans as well as prehistoric creatures. Towards the end of the summer, the expedition left the Flaming Cliffs to make its way back to China, stopping for another dig nearer to the Great Wall. Here they pitched tents and settled in for the night, only to be awoken by slithering pit vipers—the drop in temperatures as autumn came on made them seek the warmth of human cots and sleeping bags. The pit vipers were dark-coloured, thick as a man's wrist and very, very poisonous. They coiled around the legs of beds, swarmed into men's hats and boots, and dropped from tent-poles. Andrews, who was deathly afraid of snakes, stood on his cot to fight them off with a hatchet. The men killed nearly fifty of them that night before deciding to sleep in their automobiles. They soon left what they called Viper Camp, heading back into China as snow flurried down into the desert.

Returning once again to the United States, where news of the discovery of the dinosaur eggs preceded him, Andrews was for a time the most famous scientist in America. He was offered thousands of dollars by news organisations for exclusive rights to photograph the eggs and when he gave a lecture at the Museum of Natural History, four thousand people showed up for only 1400 seats.

The final expedition

Although the political situation in Mongolia continued to deteriorate, Andrews insisted on returning there again in 1925 for one more summer of exploration, this time with his largest expedition of all, consisting of forty men and 125 camels. More dinosaur eggs were found, this time alongside the fossilised skeletons of mammals, leading Andrews to postulate that dinosaurs became extinct because warm-blooded creatures had stolen their eggs—a theory since disproved.

By 1930 it was becoming harder and harder for Andrews and the Museum of Natural History to mount expeditions into China and Mongolia. Part of this was the museum's own fault—they had sold one fossilised dinosaur egg for five thousand dollars to a wealthy collector and now the Chinese seemed to think that every dinosaur egg taken from their country was worth this sum. (In fact, the selling of the egg had been a carefully orchestrated publicity stunt to raise funds for the museum, but, as Andrews later wrote, '[the Chinese] could never be made to understand that this was a purely fictitious price'.) Still, with superhuman effort, Andrews was able to take one last expedition into Mongolia. He had with him a small group that included the Jesuit geologist, palaeontologist and theologian Père Teilhard de Chardin, author of the influential book *The Phenomenon of Man*. (De Chardin had assisted in the 1929 excavation of the 500,000-year-old skeleton of Peking Man.) It was de Chardin who stumbled on a basin (such depressions in the ground were often signs of ancient bogs or quagmires where dinosaurs became caught) that would produce a stunning treasure trove of fossils, including the gigantic bones of more than a dozen mastodons. Even more amazing, they found the intact skeleton of an adult female mastodon lying on her side. Inside her pelvic cavity was the skull and jaw of an unborn mastodon.

It was a triumphant way to end the five major expeditions Andrews had led to Mongolia from 1922 to 1930. But with two groups of bandits beginning to fight an all-out war for the territory the scientists were working in, it was time to leave Mongolia for good. Andrews was to stay in China until 1932, working on his account of the expeditions. In 1935 he became director of the American Museum of Natural History, a post he held until 1941, and he would continue to be idolised by millions of Americans, the very epitome of a daring explorer, until his death in California in 1960 at the age of seventy-six.

Roy Chapman Andrews may not have been Indiana Jones, but he was, in a way, far more influential. Ever the showman—the writer Douglas Preston, in his book *Dinosaurs in the Attic*, says that Andrews 'created an image and lived it out impeccably'—he had, through his books and lectures, brought the study of dinosaurs out of its dusty obscurity and into the minds of the youth of the world—men and women who would become the next generation of dinosaur hunters.

W. DOUGLAS BURDEN
Here be dragons!

Back in the fifteenth and sixteenth centuries, when explorers first probed the vast and unknown reaches of the world in sailing vessels that could at any moment fall off the edge of the earth or find themselves hurled into the sky in apocalyptic waterspouts, mapmakers would often mark uncharted shores or empty patches of ocean with the notation *HIC SUNT DRACONES*.

This was Latin for 'Here be dragons!' and it was sometimes accompanied by an illustration of a fire-breathing dragon or a sea serpent. Even then, of course, experienced mariners did not think there were literally dragons there—merely that the land or water was unexplored or possibly treacherous. As time went on and more and more of the world became known, this quaint notation fell from use (although sea serpents were often used to illustrate corners of maps) because, after all, there were no such things as dragons.

But one day in 1910 a group of pearl drivers, searching for virgin oyster beds in the reefs off obscure islands in the Lesser Sunda Island group, in Indonesia, happened to land on Komodo Island. Although Portuguese or Japanese sailors had probably landed on Komodo before, a long time ago, no one lived in the place but a forlorn group of Malay convicts, exiled there for forgotten crimes. Fierce currents and the ravenous sharks that frequented the near waters of the island had kept most people away. When the pearl divers arrived on shore, they were astonished to find, lumbering and hissing around the rocky beaches, the very dragons medieval mapmakers had so feared. These creatures were foul-smelling, aggressive, and could run at quite a clip when provoked, and so the divers made a hasty retreat.

However, fantastic stories of these animals made their way back to P. A. Ouwens, director of the Zoological Museum in Buitenzorg, Java (Indonesia was at the time a colonial possession of the Dutch). Intrigued, Ouwens sent some brave hunters to bring back a specimen, and when the scaly-skinned, 2.7-metre (9-foot) long reptile arrived (dead, of course, for trying to apprehend the creature alive was considered suicide) Ouwens named it *Varanus komodensis* (*varanus* derives from an ancient Arabic word meaning 'lizard').

People of the early twentieth century, however, simply called the great lizard the Komodo dragon.

Dragon hunting, anyone?

About fifteen years later, in a very different locale—a penthouse apartment overlooking the west side of New York's Central Park—a twenty-eight-year-old man arrived home one evening and said to his wife: 'Care to go dragon hunting?'

The man's name was William Douglas Burden. Born in New England in 1898, Burden was quite wealthy, thanks to a family fortune, and had spent much of his time travelling the world, indulging his particular passion, which was hunting. He was not merely a mindless killer of animals, however—as a trustee of the prestigious American Museum of Natural History, he often volunteered his services in order to provide specimens of dangerous animals. He travelled the world during the 1920s, bagging Indonesian tigers, African rhinoceros, prancing Mongolian sheep and the rare red bear of the Himalayas. Stuffed, these animals adorned the halls of the museum for the edification of thousands.

After reading about the Komodo dragon, Burden had decided that he wanted to be the first man to bring back a live specimen. For many men, this would have been an idle fantasy, but Burden could afford to pay for the expedition himself, and he had the ear of Henry Fairfield Osborn, president of the Museum of Natural History. As soon as Osborn approved the idea (which meant the expedition had official museum backing, giving it a cachet that could open many doors), Burden set about picking the right people to accompany him. Aside from his adventurous wife, Katharine, a skilled photographer, he needed a big-game hunter with jungle experience, and picked his long-time hunting companion, the Frenchman F. J. Defosse. Defosse was, as Burden wrote, 'covered with scars from encounters with wild beasts of the jungle', a man who hated civilisation but who could be as stealthy and sure as a panther in the wild. Burden hired Dr E. R. Dunn of Smith College as the group's herpetologist—an expert in lizards was obviously needed—and then included a Chinese newsreel cameraman hired from the news agency Pathé News, one Lee Fai—Burden wanted to create the first live motion picture of the dragons, to be shown at the museum and elsewhere. Rounding out the group would be fifteen Malay bearers and Burden's personal manservant, Chu.

Journey to Komodo

Burden and Katharine were to meet the rest of the group in Singapore, but they did not take a direct route there. Their first stop would be Beijing, where they had decided to meet their friend, Roy Chapman Andrews, who had casually invited them 'to go for a few days up into the Gobi Desert to see the wonderful work that the Central Asiatic Expedition [Asiatic Zoological Expedition] was doing' (see page 302). They set off from San Francisco in the spring of 1926, taking an ocean liner to Japan. After a rough passage, they landed in Yokohama but then heard that civil war in China had rendered conditions there extremely difficult for the traveller. However, a cable sent to Andrews drew the one-word reply: 'Come!' Despite the danger, Burden said to his wife: 'There is plenty of time to be careful and to play it safe when we're a doddering old couple, tottering on the edge of the grave. Let's go.'

And so they did, slumming their way across the Inland Sea in a dirty Japanese freighter. Landing in China, they found a country where violent civil war raged between the factions of Nationalist leader Chiang Kai-shek and numerous warlords, and where a virulent suspicion of foreigners was endemic. They were lucky enough to meet a man who would take them into the besieged city of Peking (Beijing), where Burden chanced upon Andrews in a crowd—an encounter that, Andrews later wrote, 'seemed miraculous'. However, conditions were far too volatile for the couple to consider going to Mongolia, particularly when they had another goal in mind, and so, saying goodbye to Andrews, they made their way south, first to Hong Kong, then to Manila and finally to Singapore, where they rendezvoused with the rest of the expedition.

Burden then led the group to Batavia (currently Jakarta) in Java, where he heard the unwelcome news that the Dutch were no longer allowing visitors to Komodo. Such were his powers of persuasion (and possibly of purse) that he not only convinced them to let him go to Komodo, but he got them to allow him to cull fifteen specimens of the dragons, alive if he could get them. They even threw in an old steamer to carry the entire group—the interestingly named SS *Dog*. Thus it was that after a stop at Bali—where Burden indulged in a spot of tiger hunting and also took numerous photographs of bare-breasted Balinese beauties (which adorn his memoir *Dragon Lizards of Komodo*)—the *Dog* approached its final destination.

'A lost world'

Komodo Island, 32 kilometres long by 20 kilometres wide (20 by 12 miles), lies in the waters of the Lintah Strait, between the larger islands of Flores and Sumbawa. The Lintah Strait, the great English naturalist Alfred Russel Wallace once wrote, contains violent rip tides that make the sea 'boil and foam and dance like the rapids below a cataract, so that vessels are swept about helpless and sometimes swamped in the finest weather, and under the brightest skies'.

Braving these waters, where the sea raced as fast as thirteen knots, the expert captain of the *Dog* soon brought his vessel within sight of Komodo Island. 'With its sharp, serrated skyline, its gnarled mountains, its mellow, sunwashed valleys, and its giant [volcanic] pinnacles that bared themselves like fangs to the sky, it looked more fantastic than the mountains of the moon,' Burden wrote. 'As we drew nearer, we seemed to see a prehistoric landscape—a lost world—unfold before us ... It is a melancholy land, a fitting abode for the weird creatures that lived in the dawn of time ... a suitable haunt for the predatory dragon lizards.'

Yet when Burden landed in sun-splashed Python Bay, on the island's east end, all seemed peaceful. He and his wife strolled along the beach, admiring the wildflowers and trees of this nearly untouched island. They were astonished at how abundant game was, how deer and wild boar were evident everywhere, and how the shallows of the bay teemed with fish. Spending the night on the ship, Burden thrilled to the sound of Malay drums across the waters from nearby islands and watched as a 'golden sunset painted the rocky islets and the purple sea in glorious colors'. (He painted a less romantic picture of the tiny and perhaps understandably apathetic group of convicts he found there—'a degenerate lot of diseased people'—who included men Burden at first thought lepers, but who turned out to be albinos.)

This estimation of Komodo as an untouched tropical paradise persisted into the next day, when Burden, Defosse and Dunn went out exploring. They made their way inland, up the great volcanic peaks whose slopes were so covered with sharp lava rocks that they cut up even Burden's well-made boots. At about 600 metres (2000 feet) in altitude, they split up to search the area for a suitable camping spot. Accompanied by a few Malay bearers, Burden found a plateau that was marked by a pool of cool, limpid water. Standing on its shore, he happened to look down and was startled by what he saw—a gigantic claw mark. It was the type of mark he had seen among fossilised remains of dinosaurs, except this mark was in mud and was obviously recent. Burden realised that he had found a drinking hole of the Komodo dragon. He also discovered the heretofore unrealised presence on the island of dangerous wild water buffalo. Burden had shot a deer and was making his way down the slope when he heard a crashing in the jungle behind him. 'It sounded as though a whole bamboo forest were being broken into splinters ... I jumped back, out of the jungle, and just as I emerged into the open, I saw a great solitary bull coming full speed straight towards me. His nose was in the air, his nostrils greatly dilated, and his sweeping horns seemed to be laid back against his flanks.'

The bull was perhaps 20 metres (20 yards) away. Not having the steel-jacketed bullets needed to penetrate the hide of such a beast, Burden felt the most suitable option was flight. Racing back towards the pool of water, he climbed a high rock and waited there for his doom but, inexplicably, the water buffalo stopped, and then finally disappeared down the slope. Burden had learned a lesson that stood him in good stead on Komodo—always carry the proper munitions—for there was a lot on this island that could do him harm.

'He was a monster'

The expedition now moved off the *Dog* and set up camp on the plateau Burden had first explored. The next day, he once again set out alone to explore. Climbing higher into the mountains, he happened to look to his left and saw his first dragon. In *Dragon Lizards of Komodo*, he captures the moment:

> He was a monster—huge and hoary ... The lizard was working his way slowly down from the mountain crags. The sun slanted down the hill, so that a black shadow preceded the black beast as he came. It was a perfectly marvelous sight—a primeval monster in a primeval

*setting—sufficient to give any hunter a real thrill. Had he stood up
on his hind legs, as I now know they can, the dinosaurian picture
would have been complete. He stalked slowly and sedately along,
obviously hunting for something in the grass, his yellow tongue working
incessantly, his magnificent head swinging ponderously this way and
that ... Against a background of sunburnt grass, this particular beast
looked quite black with age, and I felt sure he bore the battle scars of
many a fierce encounter amid the deep recesses of the isle.*

Burden's depiction of the Komodo dragon as a prehistoric
creature is quite accurate. Scientists believe that the ancestors of the
creatures arose in Asia and migrated to Australia some forty million
years ago. As world sea levels dropped some four millions years ago,
the Komodo migrated to the emerging shelves of land of what is
now Indonesia. Sea levels then rose again, leaving them isolated on a
few small islands and dominating the ecosystem there.

Thrilled by the sight of his first Komodo, Burden lost no time
in beginning his hunt. First, he and Defosse shot wild deer and boar
with which to bait traps—while Komodo dragons hunt for their
meals, they prefer to eat dead, rotting meat. Then, with the corpse
tied to a stake, the hunters would hide in a *boma*, or blind, and watch
the dragons eat. Even for experienced hunters, the sight was a
disgusting one:

*In the process of gorging himself, the long sharp claws [of the
dragon] are used indiscriminately to scrape and tear with, while the thin,
recurved teeth with serrated edges are employed to rip off great chunks
of foul meat. The beast maneuvers this while seesawing back and forth
on braced legs, giving a wrench at the bait with every backward move
... When a piece of flesh has been detached, he lifts his head and gulps
down the whole slab, regardless of size. Then he licks his chops, rubs
both sides of his face on the ground as if to clean it, and lifts his
head, the better to observe the landscape. On one occasion, a lizard*

swallowed the whole hindquarters of a boar in one gulp—hoofs, legs, hams, vertebrae and all.

The venomous dragon

While the Komodo dragon is no fire-breathing dragon of old, it is definitely the nearest thing we have to it.

Mainly inhabiting the island of Komodo (but also found on neighbouring Flores and Rinca islands), the Komodo averages 2 to 3 metres (6 to 10 feet) in length, can run up to 20 kilometres (12 miles) an hour and climb trees, and can smell carrion—its favourite meal—up to 8 kilometres (5 miles) away. They can also stand on their hind legs, supporting themselves with their huge and powerful tails, which can fell a charging water buffalo with one swipe.

The Komodo is top dragon of its ecosystem, living as long as fifty years, and is not afraid of attacking any animal, even a king cobra, although it prefers carrion. Its gluttonous eating habits are astounding—dragons can consume up to eighty per cent of their body weight in one meal, although they can survive on as few as twelve large meals a year. The Komodo's sixty sharply serrated teeth are a seething mass of bacteria containing such pathogens as E. coli—it was originally thought animals wounded by the Komodo died from these virulent bacteria, but recent research has shown that the Komodo possesses venom glands and, when biting, injects poison into its victim.

Komodo Island has a current human population of some two thousand (descendants of convicts originally exiled there) and living near dragons can be difficult. The giant lizard has been known to raid human graves for bodies and, rarely, kill people—the last known fatal attack took place in March 2009, when a fruit picker on Komodo was attacked and killed by two dragons after falling out of a tree. He died within minutes of massive wounds.

Komodo dragons when they engorge themselves in this way often ram their heads against trees in order to force down their meal.

They are able to breathe during this process because of a small tube under their tongues that leads to their lungs. To further disgust the watcher, they continuously vomit indigestible and mucus-covered bones, hoofs and teeth. Rubbing their faces on the ground (observed by Burden) is a way to scrape away the thick mucus.

Katharine's close escape

Burden and Defosse shot several smaller dragons—they found that this was not a difficult task, as the animals were practically deaf, although one had to keep clear of their powerful tails and use steel-jacketed bullets—and captured a few others by placing rotting meat within a spring noose. When the Komodo nosed in for dinner, the trap would be sprung, sending the animal flying into the air, caught tight by a rope around its neck. A few more ropes would secure it and it would be brought back, caged, to the camp.

It was during an outing into the jungle with Defosse to check traps that had been set the night before that Katharine Burden experienced the closest call of the expedition. Finding one trap in which the deer used as bait had been torn from its stake, without capturing the Komodo, Defosse and Katharine set off to find the creature, each taking separate paths. Katharine heard a sound in the jungle to her right and saw a Komodo dragon come out of the jungle and pause, staring intently at the remaining half of the deer carcase. Just then, the unarmed Katharine realised that, unfortunately, she was standing directly between the dragon and the bait.

She ducked down in the tall grass and watched as the creature approached, swaying back and forth on his hind legs, 'a hoary customer, black as dead lava'. She could not decide what to do. The lizard was moving directly towards her. If she jumped up and ran, it might chase her. If she continued to lie there in the tall grass, it might go by her, intent on the rotting meat in the trap, but then again, it might not. Praying for Defosse to return, Katharine decided to stay where she was. The animal was only a few metres away and she heard 'the short hissing that came like a gust of evil wind' and watched the 'darting, snake-like tongue' that seemed to be probing the tall grass for her.

In fact, the kind of hissing or chuffing breath she describes is characteristic of a Komodo before it makes its attack, and its sense of smell (though its tongue) is its chief way of locating prey, so there is every possibility that she would soon have been ripped apart by the lizard. But just then Defosse arrived and put two bullets in the Komodo, probably saving her life.

Burden was having success in capturing small lizard specimens, killing them and preserving their bodies for further study, but he longed to bring back, alive, one of the really large dragons—as much as 3 metres (10 feet) in length and weighing over 135 kilograms (300 pounds)—that ruled the island. Since his usual traps were not strong enough to hold such a creature, he decided to build a reinforced one. One day, Defosse shot a boar and dragged it into the forest, followed by Burden and his Malay helpers. After staking the corpse of the boar to the ground, they pounded in heavy stakes all around the bait, leaving an opening at one end. Rattan or thin cane was used to lash the stakes together and then the whole thing was camouflaged with leaves and branches.

Above the trap, a small tree was stripped of most its branches and then, with fifteen men working together, was bent back and a noose was attached to its top. But instead of having the trap go off automatically when the Komodo poked its head into the noose, Burden rigged it so that it would not release unless he, hidden in a nearby blind, pulled a trip string. Then he, Defosse and Dunn settled back in the *boma* to wait. Waiting in the jungle was no easy task, since the blind was invaded by poisonous scorpions, but finally their patience was rewarded. A young Komodo strolled up to take a sniff of the bait, but it very soon made way for a larger male, which attempted with all its might to pull the carcase off the stake. But it was tied very securely and Burden decided to wait—he wanted an even larger lizard.

All of a sudden, the lizard worrying the bait put its head up, and then turned and raced off into the jungle 'as if the very devil himself were after him'. Burden cautioned the men to be silent, and within moments a huge black Komodo appeared on the edge of the clearing. This creature was obviously at the head of the Komodo pecking order. It stood there for a full half hour, staring straight at the blind, seeming to look Burden right in the eye. Then:

> *He started forward again. He was heading right for the* boma ... *I could see the ugly brute very well. He looked black as ink. His bony armor was scarred and blistered. His eyes, deep set in their sockets, looked out at the world from underneath overhanging brows ... Now the creature's footsteps were audible.*
>
> *He passed right by one side of the* boma. *I could have reached out and touched him with my hand, and I had the tingling sensation of having a dragon walk by within a yard of where I was standing.*

The dragon finally walked by the blind and arrived at the trap, but it seemed extremely suspicious of the whole set-up. It kept walking up to the opening and putting its head part way into the noose, but not far enough for Burden to spring the trap. The animal would stop what it was doing, examine the surrounding jungle for minutes at a time, and then partly put its head in the noose. It repeated this procedure over and over again, to the consternation of Burden and the men in the *boma*. Then, finally, when Burden had given it up as a lost cause, the animal appeared to make up its mind. It placed its head fully inside the noose and began eating, at which point Burden sprang the trap.

The huge animal suddenly flew skywards. For a moment the surprised dragon jerked wildly in the air, but the tree could not support its weight and partially broke. On the ground again, the giant lizard thrashed in a frenzy, trying to pull the noose off. It was at this point that Defosse calmly stepped into action. Leaving the blind with a coiled rope, he attempted to lasso the dragon—surely the first time in history anyone had tried such a dangerous stunt. He missed the first time but was successful in flipping the noose around the dragon's neck on his second throw. He then tied the rope to a tree and three more ropes were attached to the wildly struggling lizard, which had begun, in its distress, to vomit a foul-smelling substance. Finally, though, with the help of their Malay bearers, the men were able to lash the creature to a pole and bring it back to camp.

Once in camp, they put the lizard into a cage—made of heavy timber and lashed together with wire and steel mesh—they had built especially for it. They congratulated themselves on a job well done, but then the dragon, freed inside its cage, began lashing around with its tail and vomiting continuously. The smell was so foul that Burden ordered that the cage be dragged a quarter of a mile (500 metres) from camp.

In the morning, he and the rest went out to examine their prize find, only to discover that the dragon, with its extraordinarily powerful jaws, had torn through the heavy steel mesh at the top of the cage and disappeared into the night.

'The lizards, I'm sorry to say'

Despite his disappointment over the one that got away, Burden felt fairly satisfied with his trip when, about a month later, the expedition pulled up stakes and left Komodo Island forever on the trusty *Dog*. Komodo was soon just a 'shadow that loomed up weird and indefinite in the distance', but Burden had on board with him twelve dead and preserved Komodos and two live ones, securely kept in iron cages. Back in New York, Burden was greeted with great excitement. Some of the twelve dead specimens were parcelled out to scientists for study, while others were stuffed and placed in a diorama at the American Museum of Natural History—they can still be seen there today, eating a dead boar.

As for the two live dragons? Well, they were immediately taken to the Bronx Zoo, where they became an overnight attraction, but soon the inadvisability of taking such a creature from its natural habitat was evident, since the dragons died shortly afterwards. Burden noted that 'it was painful to see the broken-spirited beasts that barely had strength to drag themselves from one end of the cage to another'. Perhaps so, but Burden was not so pained (perhaps he was still holding a grudge over the one that got away) that he did not print in *Dragon Lizards of Komodo* a little ditty, supposedly composed by one of his friends:

> *But the lizards, I'm sorry to say,*
> *Became gaunt, and just faded away;*
> *For they could only thrive*
> *Eating chickens alive,*
> *Which was banned by the S.P.C.A.*

William Douglas Burden went on to a varied career, doing more big-game hunting but also dabbling in film making. His 1930 docudrama *The Silent Enemy*, about the fate of Canada's Ojibwa Indians—an epic that was originally six hours long—was hailed by critics as 'superb', although Burden had the misfortune to release a silent film just as talkies were coming in. (Burden also became a footnote in film history himself when his journey to Komodo Island served as an inspiration for the movie *King Kong*.)

The Komodo dragon is now protected under Indonesian national law and part of the island has been put aside as the Komodo National Park, so that the big lizards no longer have to worry about taking a bite of a nice smelly boar and winding up sailing skywards, ropes around their necks. However, modern society has found a way to co-opt the dragon. Enter any Starbucks Coffee Shop and browse the packages of coffee for sale. There you'll find Komodo Dragon Blend: 'Spicy, herbal and earthy, each sip evokes the lushness of Indonesia.'

Kong and Komodo

If the strange landscape of remote Komodo Island—its 'sharp serrated skyline' and the 'giant pinnacles that bared themselves like fangs to the sky'—strike a chord in your psyche, it may be because they remind you of the menacing terrain of Skull Island in the classic 1933 movie King Kong.

There is a reason for that. The movie's creator, Merian C. Cooper, was fascinated by the story of Douglas Burden's trip to Komodo Island and wrote to him several times asking for details, especially when it came to the island's terrain and the sad fate of the Komodos that could not survive life in a New York zoo.

In 1964 Cooper and Burden had this exchange of reminiscing letters. Burden wrote: 'I remember, for example, that you were quite intrigued by my description of prehistory Komodo Island and the dragon lizards that inhabited it ... You especially liked the strength of words beginning with 'K.' such as Kodak, Kodiak Island and Komodo. It was then I think that you came up with the idea of Kong as a possible title for a gorilla picture.'

And Cooper replied: 'When you told me that the two Komodo Dragons you brought back to the Bronx Zoo, where they drew great crowds, were eventually killed by civilization, I immediately thought of doing the same thing with my Giant Gorilla ...'

The gallant, intrepid hunter of King Kong *reminds one of Burden (though there were numerous such types at the time—Roy Chapman Andrews was another) and his lovely wife, Katharine (saved from the jaws of the Komodo at the last minute by Defosse) was perhaps also an inspiration for the part played by Fay Wray—the woman in peril from the savage beast.*

BIBLIOGRAPHY

Andrews, Roy Chapman. *On the Trail of Ancient Man*. G.P. Putnam's Sons, New York, 1926

Andrews, Roy Chapman. *Under a Lucky Star*. Viking Press, New York, 1943

Arnold, Ann. *Sea Cows, Shamans, and Scurvy: Alaska's First Naturalist: Georg Wilhelm Steller*. Farrar Straus Giroux, New York, 2008

Banks, Joseph. *The Endeavor Journal of Sir Joseph Banks*. Project Gutenberg Australia: http://gutenberg.net.au/ebooks05/0501141h.html

Bartram, William. *Travels and Other Writings*. The Library of America, New York, 1996

Bartram Heritage Report. http://www.bartramtrail.org/pages/frame1.html

Bates, Henry Walter. *The Naturalist on the River Amazons: The Search for Evolution*. The Narrative Press, Santa Barbara, California, 2002

Beebe, William. *Half-Mile Down*. Duell, Sloan and Pearce, New York, 1934

Blunt, Wilfrid. *The Compleat Naturalist: A Life of Linnaeus*. The Viking Press, New York, 1971

Bown, Stephen R. *The Naturalists: Scientific Travelers in the Golden Age of Natural History*. Barnes & Noble, New York, 2002

Brown, Jane. *Tales of the Rose Tree: Ravishing Rhododendrons and Their Travels Around the World*. David R. Godine, New York, 2006

Bull, Bartle. *Safari: A Chronicle of Adventure*. De Capo Press, New York, 2006

Cashin, Edward J. *William Bartram and the American Revolution on the Southern Frontier*. University of South Carolina Press, Columbia, South Carolina, 2000

Coats, Alice M. *The Plant Hunters*. McGraw-Hill Books, New York, 1969

Cummins, Joseph. *History's Great Untold Stories: Larger Than Life Characters*

& *Dramatic Events That Changed the World*. Murdoch Books, Sydney, 2006

Dodge, Bertha S. *It Started in Eden: How the Plant-Hunters and the Plants They Found Changed the Course of History*. McGraw-Hill Books, New York, 1979

Douglas, David. *Journal Kept by David Douglas During His Travels in North America*. The Royal Horticultural Society, London, 1914

Edington, Brian. *Charles Waterton: A Biography*. Lutterworth Press, London, 1997

Emling, Shelley. *The Fossil Hunter. Dinosaurs, Evolution, and the Woman Whose Discoveries Changed the World*. St. Martin's Press, New York, 2009

Fleming, Fergus. *Off the Map: Tales of Endurance and Exploration*. Grove Press, New York, 2004

Forrest, George. *The Journeys and Plant Introductions of George Forrest*. Oxford University Press for Royal Horticultural Society, London, 1952

Freedman, Russell. *An Indian Winter*. Holiday House, New York, 1992

Fullerton, Patricia. *The Flower Hunter: Ellis Rowan*. The National Library of Australia, Sydney, 2002

Gallenkamp, Charles. *Dragon Hunter: Roy Chapman Andrews and the Central Asian Expeditions*. Penguin Books, New York, 2002

Golder, F.A. (ed.). *Steller's Journal of the Sea Voyage from Kamchatka to America and Return on the Second Expedition*. American Geographical Society, New York, 1925

Hood, Robert. *To the Arctic by Canoe, 1891-1821: The Journal and Paintings of Robert Hood*. McGill-Queens University Press, Montreal, 1994. Copyright © 1994 by Robert Hood. Quote on page 157 reprinted with permission of McGill-Queens Press

Hooker, Joseph Dalton. *Himalayan Journals or, Notes of A Naturalist*. John Murray, London, 1854

Houston, C. Stuart. *Arctic Ordeal: The Journal of John Richardson, Surgeon-Naturalist with Franklin, 1820–1822*. McGill-Queens University Press, Montreal, 1984

Humboldt, Alexander von. *Personal Narratives of Travels to the Equinoctial Regions of America, 1799–1804,* vols 1–5. Project Gutenberg http://www.gutenberg.org/dirs/etext04/qnct210.txt

Huxley, Robert (ed.). *The Great Naturalists*. Thames & Hudson, London, 2007

Kingsley, Mary H. *Travels in West Africa: Congo Français, Corisco and Cameroons*. Macmillan & Co. Ltd, London, 1904

Lindsay, Ann & Syd House. *The Tree Collector: The Life and Explorations of David Douglas*. Aurum Press, London, 2005

Linnaeus, Carl. *A Tour in Lapland*. White & Cochran, London, 1811

Maximilian, Prince of Wied. *Travels in the Interior of North America,* vols 1–2. The Arthur H. Clark Company, Cleveland, Ohio, 1906

Millais, J. G. *Life of Frederick Courteney Selous, D.S.O.* Longmans, Green & Co., New York, 1919

Musgrave, Toby, Chris Gardner & Will Musgrave. *The Plant Hunters: Two Hundred Years of Adventure and Discovery Around the World*. Seven Dials Press, London, 2000

North, Marianne. *Recollections of a Happy Life,* vols 1–2. Macmillan & Co., London, 1892

O'Brien, Patrick. *Joseph Banks: A Life*. The University of Chicago Press, Chicago, 1987

Potter, Jennifer. *Strange Blooms: The Curious Lives and Adventures of the John Tradescants*. Atlantic Books, London, 2006

Rafinesque, Samuel Constantine. *A Life of Travels in North America and South Europe*. Published by the author, Philadelphia, 1836

Robbins, Peggy. 'The Oddest of Characters', *American Heritage Magazine*, June/July 1985

Selous, Frederick Courteney. *A Hunter's Wanderings in Africa: Being a Narrative of Nine Years Spent among the Game of the Far Interior of South Africa*. Macmillan & Co., London, 1907

Sheffield, Susan Le-May. *Revealing New Worlds: Three Victorian Women Naturalists*. Routledge, London, 2001. Copyright © 2001 by Susan Le-May Sheffield. Quote on page 234 reprinted with permission of Routledge

Shipman, Pat. *The Man Who Found the Missing Link: Eugene Dubois and His Lifelong Quest to Prove Darwin Right*. Harvard University Press, Cambridge, Mass., 2001

Spalding, David A. E. *Dinosaur Hunters: Eccentric Amateurs and Obsessed Professionals.* Prima Publishing, Rockland, California, 1993

Todd, Kim. *Chrysalis: Maria Sibylla Merian and the Secrets of Metamorphosis.* Harcourt Inc., New York, 2007

Warren, Leonard. *Constantine Samuel Rafinesque: A Voice in the American Wilderness.* The University of Kentucky Press, Louisville, Kentucky, 2004

Illustrations

Cover, *Tridacna gigas;* The John Tradescants, page 9: *Tilia grandiflora*, lime tree; Maria Sibylla Merian, page 22: *Papilionida*; Carl Linnaeus, page 36: *Anagallis arvensis*; Georg Steller, page 51: *Gallurus glandarius*; William Bartram, page 66: *Franklinia altamaha;* Joseph Banks, page 79: *Banksia serrata;* Alexander von Humboldt, page 95: *Anguilla vulgaris*; Charles Waterton, page 109: *Strychnos toxiflora*; Prince Maximilian of Wied, page 124: *Quercus tinctorial*; Constantine Samuel Rafinesque, page 138: *Zeus faber*, dory (top left), *Lophius piscatorius*, wide gap (bottom right); John Richardson, page 152: *Lota vulgaris*, burbot; David Douglas, page 168: *Dodecatheon integrifolium*; Mary Anning, page 183: *Plesiosaurus macrocephalus*; Sir Joseph Dalton Hooker, page 198: *Rhododendron dalhousiae*; Henry Walter Bates, page 213: *Ramphastus tucanus*; Marianne North, page 228: *Rosa moscata*; Frederick Courteney Selous, page 244: *Felis leo*; Eugène Dubois, page 257: *Pithecanthropus erectus* (Java Man); Mary Kingsley, page 270: *Felis leopardus*; George Forrest, page 286: *Bambusa arundinacea*; Roy Chapman Andrews, page 302: fossil of the tertiary period; W. Douglas Burden, page 320: *Draco dandini.*

INDEX

Published in 2010 by Pier 9, an imprint of Murdoch Books Pty Limited

Murdoch Books Australia
Pier 8/9
23 Hickson Road
Millers Point NSW 2000
Phone: +61 (0) 2 8220 2000
Fax: +61 (0) 2 8220 2558
www.murdochbooks.com.au

Murdoch Books UK Limited
Erico House, 6th Floor
93–99 Upper Richmond Road
Putney, London SW15 2TG
Phone: +44 (0) 20 8785 5995
Fax: +44 (0) 20 8785 5985
www.murdochbooks.co.uk

Publisher: Diana Hill
Project Editor: Sophia Oravecz
Designer: Jay Ryves at Future Classic

National Library of Australia Cataloguing-in-Publication entry
Author: Cummins, Joseph.
Title: Eaten by a giant clam : great adventures in natural science /
                  Joseph Cummins.
ISBN: 978-1-74196-753-1 (hbk.)
Notes: Includes bibliographical references and index.
Subjects: Natural history.
                  Science.
Dewey Number: 508

A catalogue record for this book is available from the British Library.

Printed by Hang Tai Printing Company Limited, China.